Remote Control of War Dogs

By
John J. Romba

Army Land Warfare Laboratory

Fredonia Books
Amsterdam, The Netherlands

Remote Control of War Dogs

by
John J. Romba
Army Land Warfare Laboratory

ISBN: 1-4101-0722-1

Copyright © 2004 by Fredonia Books

Reprinted from the 1974 edition

Fredonia Books
Amsterdam, The Netherlands
http://www.fredoniabooks.com

All rights reserved, including the right to reproduce this book, or portions thereof, in any form.

PREFACE

The work described in this report was performed under LWL Tasks 08-B-72, Remote Control of War Dogs, and 03-B-74, Remotely Controlled Scout Dog. I gratefully acknowledge the help of all of the people who worked on the project, who were too numerous to mention individually by name. Those persons whose contributions to the program were especially significant included the following personnel of the Military Dog Detachment, 29th Infantry Brigade, Ft. Benning, GA, formerly the Military Dog Committee, USA Infantry School, Ft. Benning, GA: LTC William H. Jarvis, former Chairman, Military Dog Committee, MSG Reinard Booth, Senior Trainer, CPT Woodrow Quinn, former Commanding Officer, Military Dog Detachment, and SGTs Norbert Berriment, Emanuel Fuller, Donnie Jones, Charles Carrier and Robert Williams. I am also deeply obligated to COL M. W. Castleberry, Director, Biosensor Research Team, Walter Reed Army Institute of Research, at Edgewood Arsenal, MD, who unstintingly provided support and facilities without which it would have been impossible to carry out the investigation of automated training procedures. I am grateful to COL Castleberry for his many invaluable comments and suggestions, and SP/4 Steven Lindsay of the Biosensor Research Team. I also wish to thank Dr. Max Krauss, Chief, Biological Sciences Branch, LWL, for his guidance and advice during the course of this program.

TABLE OF CONTENTS

	PAGE
PREFACE	v
LIST OF ILLUSTRATIONS	xi

I. INTRODUCTION — 1

II. REINFORCING STIMULI — 4

Conditioning the Stimulus GOOD — 6
Step 1. Sampling the Food Reinforcer — 6
Step 2. Pairing the Word GOOD with Food — 6
Step 3. Test of Stimulus Conditioning — 6

The Reinforcing Stimulus OUT — 7
Step 1. Sampling the Food Reinforcer — 7
Step 2. Pairing the Command OUT with Food — 7
Step 3. Pairing OUT with Ball Play — 7

Conditioning the Aversive Stimulus TIME — 8

III. CONDITIONING ACCESSORY BEHAVIORS — 9

Controlled Down Response — 9
Step 1. Adaptation to Training Setting — 9
Step 2. Escape from an Aversive Stimulus — 9
Step 3. Learning to Anticipate the Aversive Stimulus — 10
Step 4. DOWN-STAY Training — 10
Step 5. Pairing a Tone with Tug-On-Leash — 10
Step 6. Change in Trainer Stance — 11
Step 7. Rise Training — 11
Step 8. Conditioning RISE-STAY — 11
Step 9. Transfer to Other Environmental Settings — 11

Controlled Recall Response — 11
Step 1. Pairing the Voiced COME and GOOD Stimuli — 12
Step 2. Discrimination Training to the Word COME — 12
Step 3. Pairing a Tone Signal and the Verbal Command COME — 12

IV. DETECTION INDICATOR RESPONSES — 13

Controlled Halt-Stand Stay — 13
Step 1. Conditioning the Halt Command, HUT and STAY — 13
Step 2. Training the Dog to Stand-Stay Away from Trainer — 13
Step 3. Transfer of Detection Indicator Response to People in View — 14
Step 4. Transfer of Detection Response to People in Concealment — 15

TABLE OF CONTENTS (Cont'd)

		PAGE
Controlled Sit Response		16
Step 1. Forcing the Sit Response		16
Step 2. Pairing a Touch Stimulus with the Pressure Stimulus		16
Step 3. Pairing the Verbal SIT with the Touch Stimulus		17
Step 4. The Transfer Stimulus		17
Step 5. Adaptation to the Work Station		17
Step 6. Shaping Head Lowering		19
Step 7. Shaping a Nose Touch Response to the Transfer Stimulus		19
Step 8. The Transfer Stimulus Card Position is Varied		19
Step 9. The Trainer Assumes a Standing Position		19
Step 10. Sit Response to the Transfer Card		19
Step 11. Pairing the Transfer Stimulus with Other Objects		20

V. THE SEARCH CHAIN 21

    Step 1. Stimulus Conditioning — 21
    Step 2. Shaping Moving Toward the Trainer — 22
    Step 3. Shaping Movement Around the Trainer — 22
    Step 4. Stop-At-Heel Training — 22
    Step 5. Stay and Move-From-Heel Training — 23
    Step 6. Adding a Detection Response to the Search Chain — 23
    Step 7. Transfer of Search Chain to Other Settings — 23

VI. CONTROL OF OUTGOING BEHAVIOR IN A FREE FIELD — 25

    Shaping Long Distance Runs — 25
    Control of Movement in Open Fields — 26
    Learning to Overcome Obstacles — 27

VII. REMOTE CONTROL — 28

    First Experiment. Exploratory Short Method — 28
    Step 1. Adaptation to the Cross — 28
    Step 2. Two-Position Discrimination — 28

    Second Experiment. Manual Training Procedures — 30

    Basic Stimulus Conditioning — 30
    Step 1. Conditioning Pattern Discrimination — 30
    Step 2. Transfer of Pattern Discrimination to Excursions in a Runway — 31
    Step 3. Learning Light-Dark Discrimination — 31
    Step 4. Reducing Reaction Time to Light Onset — 33

TABLE OF CONTENTS (Cont'd)

	PAGE
Transfer of Training to a 2-Choice Construction	33
Step 5. Escape Conditioning to the Flashing Light	33
Step 6. Introduction of an Omnidirectional Flashing Light	33
Step 7. Transfer of Direction Control from Light to Tone	35
Transfer of Training to a Cross	35
Step 8. Learning When to Stop Changing Direction	35
Third Experiment. Automation of Training Procedures	36
Conditioning Stimuli by Pavlovian Methods	36
Step 1. Adaptation and Feeder Training	38
Step 2. Feeder Stimuli made Contingent on a Response	38
Step 3. Light Discrimination Training	38
Conditioning to Light by Having it Contingent on a Response	39
Step 1. Adaptation and Center-Key Responding	39
Step 2. Shaping a Key Touch Response at the End Panels	40
Step 3. Alternating Reinforcement Schedules on Two Keys	40
Step 4. Introduction of a Time-Out Period	40
Step 5. Control of Movement by the Light Stimulus	40
Transfer of Training to a 2-Choice Construction	40
Step 6. Control of Direction by Light in the Y	41
Step 7. A New Stimulus that Defines the Interval Between Trials	41
Step 8. Non-Discrimination Trials	41
Step 9. Introduction of Flashing Light Delay	41
Step 10. Flashing Light Signal made Contingent on Entry into Wrong Alley	41
Step 11. Pairing Tone and Flashing Light	42
Transfer of Training to a Cross	42
Step 12. Single Alley Runs on the Cross	42
Step 13. Direction Control in a 2-Alley Run	43
Step 14. Direction Control in a 3-Alley Run	43
VIII. GUIDANCE CONTROL IN A FREE FIELD	44
CONCLUSIONS	46
RECOMMENDATIONS	47
DISTRIBUTION LIST	49

LIST OF ILLUSTRATIONS

Figure		Page
1	Transfer stimulus card	18
2	Cross-shaped apparatus	29
3	Straight alley apparatus	32
4	Y-shaped apparatus	34
5	Skinner box environment	37

I. INTRODUCTION

Almost a decade ago the US Army Land Warfare Laboratory (LWL) demonstrated that an Infantry scout dog can be worked reliably in the off-leash mode, at distances of 50 to 150 meters from the handler. The present study had three major objectives: (1) to develop and apply procedures suitable for training dogs to respond appropriately to radio-transmitted guidance signals at distances as great as half a mile from the handler; (2) to demonstrate the feasibility of the remote control concept in a context of potential military significance; (3) to develop a radio-transmitter-receiver system as the basis for a practical, operational system.

This report deals mainly with the effort to develop suitable training procedures. The essential feasibility of the concept was shown as training progressed and it could be demonstrated that, in fact, dogs can be remotely directed over considerable distances from the handler. The development of a radio transmitter-receiver system is described in a separate report.[1] A simplified version of the radio system, based on a reassessment of the minimal guidance input needed for the majority of practicable tactical scenarios, was subsequently developed. This version consisted of a small, hand-held tone modulator unit connected to the X-receptacle of an AN/PRC-77 radio, that functioned by means of solid-state relaxation oscillators to transmit three tones: a 400 Hz sawtooth tone (change direction), an 800 Hz tone with a 50% duty cycle (down-stay), and a 2000 Hz warbling tone (recall).[2] The radio receiver carried by the dog was an AN/PRR-9 "squad radio" receiver which was strapped to the dog's harness. Several tone-modulator units were supplied to the Military Dog Detachment, attached to the 29th Infantry Brigade, at Ft. Benning, GA. In tests at Ft. Benning, it was found that the tone-modulator-PRC-77/PRR-9 transmission-receiver system has a useful range of 2 miles. The PRC-77 with the tone modulator connected, can still be used for normal transmission-reception.

The general problem of training a dog to respond reliably to tones as guidance signals and to perform a useful scouting/detection function at considerable distances from the handler, can be structured in terms of a short-range and a long-range model. An assumption, basic to both models, is that the dog operates under primary guidance control of terrain stimuli, with control often shifting from one terrain feature to another. At any given moment, the dog may use a trail, a road, a brush line, or any directional surface configuration to guide its travel, or it may be attracted toward a distant promiment landmark, such as a tree line, a single large tree, a vehicle, a building, etc.; or its guidance may even come from moving away from something, its handler, for instance. In the short distance model, control by terrain stimuli can easily be overridden by arm and hand change-of-course signals

---

[1] Electronic Dog Handler System, LWL Technical Report No. LWL-CR-08B72, June 1973.

[2] Designed and fabricated by the Bendix Corp., Baltimore, MD under a contract monitored by the Communication and Electronics Branch of LWL.

given by the handler. The signals tell the dog to change course and they also give the new direction of travel. Change-of-course control by a tone signal in the long-distance model is more difficult to obtain. In this case, onset of the signal only tells the dog to begin turning. Cessation of the signal tells the dog to stop turning and to proceed ahead in the direction it is pointed. The change direction signal in this work was the 400 Hz sawtooth tone previously noted. At the onset of the tone, a dog was required to start turning, but it was allowed to turn either left or right. At the instant the tone went off, the dog was required to stop turning and to begin moving ahead in the new direction. Thus, a central training problem in developing a remote control capability, was to train the dog to respond reliably and precisely to the change-direction signal. The "down-stay" and "recall" signals were necessary to complete the remote capability, but the training problems involved were not as complex or difficult as in the case of the change-direction signal.

In general, the training method was based on the principles of operant conditioning.[3] The procedures that were developed emphasized positive reinforcement such as food and play, as a conditioning mechanism. Some of the conditioning process, however, did make use of negative reinforcement and punishment. Aversive consequences consisted of the withholding of positive reinforcement, removal of the animal from the work setting or increasing the amount of work required per unit of positive reinforcer. Underlying the use f these aversive events was the assumption that a dog will behave preferentially in such a way as to maintain a rewarding situation. The method permits behavior to operate freely on the environment and thereby produce changes in it. The changes that are produced are called reinforcers - these are the consequences of behavior. The proper role of the trainer is to arrange for reinforcing events to follow particular responses and not to follow other responses. By making a reinforcing event (stimulus) contingent on a particular response, the probability that that response will recur is increased. In preparing a training program the trainer not only arranges for a reinforcing event or stimulus to follow a particular response but he also defines the parameters that will determine when reinforcement of the response will be in effect. For example the command SIT defines an occasion when sitting will be followed by a morsel of food; sitting in the absence of this signal will not be reinforced. An animal thus obtains information about the nature of the consequence and its availability before the response is made. The parameters of a situation that determine whether reinforcement of a particular response will occur are called discriminative stimuli.

As a practical training matter, the trainer cannot simply wait for a specific response to occur and then reinforce it - the probability that such a response will occur within the time limits of a training session may be infinitesimal. What the trainer does is to select for reinforcement an action or a behavior that occurs with some reasonable probability and that in some way contributes

---

[3]Reynolds, G. S., A Primer of Operant Conditioning, Scott, Foresman and Co., 1968.

to the pattern of the desired end-behavior. The initial action (response) will then be emitted at a higher rate. Reinforcement now is withheld until a response more closely approximating the desired end-behavior occurs. In this way successive reinforcement-extinction steps obtain a closer approximation of the terminal behavior. By making each successive step small, the frequency of reinforcement can be maximized and aversive consequences of being wrong are minimized. The trainer can speed up this so-called shaping process by controlling the training environment in such a way that the probability that particular responses will occur is increased.

The complex behavioral activities of a remotely-controlled scout dog can be described in terms of an orderly sequence or chain of behaviors under the control of multiple stimuli. A behavior chain begins with the presentation of a discriminative stimulus. When the animal makes the required response in the presence of that stimulus a conditioned reinforcer appears (is presented). The conditioned reinforcer also acts as a discriminative stimulus for the next appropriate response. This response in turn is reinforced with another conditioned reinforcer that is also a discriminative stimulus for the next response, and so on. The last stimulus in the chain is a primary reinforcer such as food. A single primary reinforcement thus may control a long behavior sequence.

The procedures that are described in the following sections constitute a training program that is quite flexible. The accessory behaviors, detection indicator responses and the search chain can be taken up in any order and concurrently in any combination. The search chain however, is the nucleus of an operational system in which the other behaviors are integrated. Because of its inherent difficulty, developing control of the change-direction behavior is presented as a separate section in a context of three experimental sub-programs. The third of these sub-programs, or experiments, consisted of an effort to automate the training procedures leading to change-direction control.

II. REINFORCING STIMULI

Behaviors tend to be repeated when they are followed by positive reinforcers or when they avert the occurrence of negative reinforcers. Food, ball play, petting, a rawhide bone, a period of exploration are all effective positive reinforcers for dogs. They can be used as pay-off for specific kinds and amounts of work output. Behavior learned with the use of positive reinforcers can usually be improved by the delivery of negative reinforcers whenever the desired responses are not forthcoming.

Some variables that affect performance when food is used to reinforce behavior are: the dog's state of hunger (duration of deprivation), palatability and amount of the food offered, frequency of reinforcement, the emotional state of the dog, its physical condition, including its present weight relative to its free-feeding weight, and the concurrent presence of other, uncontrolled reinforcers. It was shown by the author that a good motivational level can be maintained with a dog without reducing the dog's weight or depriving it of food longer than 18 hours, provided that other conditions are met. Variation in the kind and amount of food and in delivery schedules, especially, influence performance.

Sustained work can be maintained by a dog if the food payoff per unit of work is not made too small or if the amount given at every reinforcement opportunity is not made too large. Several reinforcement schedules, operating concurrently, were found to provide the sustained work output required for learning the skills described in this report. A food pellet, often used in laboratory studies, or a semi-soft commercial dog food in cube form were delivered on a continuous reinforcement schedule, i.e., food followed every response. On the last trial of a work session, a mixture of canned beef gravy and 4 to 12 food pieces (the amount depended on the difficulty of the task) followed the required response. On the last correct response of the day, after one or more sessions, the remainder of the day's ration was given, which consisted of a mixture of any palatable semi-soft or dry commercial dog food and canned beef with gravy.

The small food pieces which were delivered on a continuous schedule, appeared to function partially as token reinforcement--information stimuli that promised the imminent appearance of something better. Good performance could not be consistently maintained solely on this continuous schedule. Petting and praise when used as token reinforcement, promising a filled food pan at the end of the work day proved, also, not to be an effective schedule combination.

Petting, like food, is an unconditioned stimulus. Normally, dogs show overt social responsiveness toward persons who handle them, even toward those who employ harsh corrections in their training. Running and jumping about with brushing contact against the person are frequently observed when dogs are petted especially during the first moments following periods of physical and social separation. In obedience training, petting and praise are often used to follow the same response that avoids an aversive correction. When used in this way, petting and praise appear to aid learning, but their role in the learning process is unclear. In a recent exploratory study, petting and praise, when used alone, failed to increase the frequency of the sit response from the level observed before conditioning. When a food reinforcer alone followed the response, the frequency of sitting increased significantly. A change back to the petting and praise reinforcement produced a decrease in responding to the pre-conditioning level. A further alternation of reinforcers again showed the conditioning and extinction effect. It plainly appears that the correction procedure and petting and praise facilitate learning when both are used together and retard learning when each is used alone. If petting and praise are effective in reducing the side effects of aversive stimuli, i.e., fear, anxiety and general behavior suppression, then they would not necessarily have to be response contingent--they could be given at any time the dog appeared to become tense or nervous. In the present study, petting and praise were not response dependent but were used mainly to energize the dogs into greater general activity during work sessions.

Even though positive reinforcers have a powerful effect on behavior, and complex repertoires can be built with them, there is no assurance that a particular positive reinforcing stimulus at the moment of a response will strengthen that response. There are gaps in our knowledge concerning the use of positive reinforcers and how to make them work well every time. Two related aversive control techniques which have been little used with dogs, but appear potentially effective in some learning situations when used with positive control methods, are "time-out from positive reinforcement" and "change in response cost." Different aversive effects are associated with time-out, of which two are readily identified. They are withdrawal from a rewarding situation and social deprivation. Response-cost simply means that the amount of work required for a unit of reinforcer is decreased when performance is good and increased when performance is poor. As aversive stimuli, they can be given when a desired response is not forthcoming or in the presence of an undesired one. They are unique in that they do not elicit fear and anxiety in an animal, as do most aversive stimuli, but they have been known to induce anger and aggression.

When a positive reinforcing stimulus is made contingent on the occurrence of a response, it increases the probability that the response will occur again. Food, ball play, and petting are examples of positive reinforcers, but they do not directly reinforce most behaviors. Consider the situation in which a trainer tries to deliver food at the exact moment of the sit response. The trainer cannot avoid tipping the animal off prior to sitting that food is imminent. Thus a competing food-begging response, behavior preparatory to consummation, is emitted prematurely. Problems of this sort can be prevented if conditioned stimuli, such as GOOD, are used to reinforce target behaviors directly, and unconditioned stimuli, such as food, follow only the consummatory behaviors. Thus, the conditioned reinforcer not only strengthens the response it follows, but sets the occasion when begging will be reinforced with food. The trainer in this way eliminates all consummatory-related behaviors during a training period until the food-associated conditioned reinforcer is delivered.

The behavior learning paradigm specifies that responses can be strengthened if they produce or avert certain stimulus consequences. It says nothing about how stimuli come to have reinforcing properties or discriminative functions. Almost all reinforcing and discriminative stimuli, it turns out, must be learned, but the learning process apparently differs from that of behavior learning. Pavlov, the Russian physiologist and experimental psychologist, developed the conditioned reflex concept to explain stimulus conditioning. Respondents, which comprise the behavior class found in the reflex, are essentially involuntary. Food, which is the principle unconditioned stimulus used in the present study, normally elicits unconditioned reflexes such as salivation and other autonomic responses in the hungry dog. The same autonomic respondents will be elicited by a previously neutral stimulus (conditioned reflex), if it is paired with the food stimulus. It appears that a necessary and sufficient condition for conditioning a previously neutral stimulus is for that stimulus to appear within one or two seconds before an unconditioned stimulus.

Procedures are given in this section that result in learning the conditioned reinforcers GOOD and OUT, and the conditioned aversive stimulus TIME. The stimulus GOOD is the principle conditioned reinforcer used for the acquisition of target behaviors. Therefore, it must be available before any response conditioning can commence. Conditioning the stimulus OUT, a more attractive reinforcer, can be done after response conditioning has begun. The aversive stimulus TIME is used mainly to sharpen the performance of an already learned behavior--the learning of this stimulus can be delayed several weeks.

Conditioning the Stimulus GOOD.

Step 1. Sampling the Food Reinforcer. The dog becomes acquainted with the food that will be used in training. The trainer puts a bit of food into the dog's mouth. Often, the food will be spit out, investigated while on the ground, and, finally, eaten. The next offering usually is readily accepted. Portions of food are presented about every three seconds, until begging behavior appears.

Step 2. Pairing the Word GOOD with Food. The stimulus sequence in the conditioning procedure is as follows: the word GOOD is enunciated first (it is not dependent on any response), followed closely by hand movement out of the food bad, followed immediately by hand opening and exposing the food, then lowering the hand to where the dog can reach the food. All this is done in a smooth continuous motion, in a way that does not startle the animal. The sequence is repeated at intervals of 3 to 5 seconds. Conditioning of the stimulus GOOD is usually accomplished in 6 to 12 sequences (trials). Since begging is never reinforced with food before GOOD is spoken, this preconsummatory behavior then will not appear until it is signalled to do so by the stimulus GOOD.

Step 3. Test of Stimulus Conditioning. Conditioning of the stimulus GOOD can be tested simply by trying to use it to acquire an easy-to-learn response, such as movement away from the trainer. The stimulus GOOD is spoken every time the dog turns and moves away. If the procedure results in an increase in the frequency of this behavior, then the word GOOD is ready for use as a conditioned reinforcer to produce learning and to maintain acquired skills.

The Reinforcing Stimulus, OUT.

The word OUT is used to strengthen behaviors by following them closely in time OUT apparently is a more powerful reinforcer than GOOD because it promises the dog a better payoff. Thus, OUT is associated with various primary reinforcers such as large, tasty amounts of food, ball play, or a dip in the pool on a hot day. OUT is given in place of GOOD after a moderate amount of work has been accomplished or a complex task has been performed. It is usually reserved for the last trial of a session.

Step 1. Sampling the Food Reinforcer. A substantial amount of the dog's daily food ration is given on the last training trial of the day. This should consist of a quantity of semi-moist packaged dog food mixed with canned dog food and gravy. As with any other kind of food, sampling is done to determine how well the dog likes it. If the dog eagerly eats the food, it is ready for conditioning to the signal OUT. If the dog merely nibbles at the food and eats sparingly, it can be made more hungry the next time, or a more desirable food mixture can be found.

Step 2. Pairing the Command OUT with Food. In this step, the dog is taught that a particularly palatable portion of food regularly follows the word OUT. The dog is tied to a 6 ft. stake out line. The part of its food ration that was not consumed in the training sessions is placed about 2 ft. beyond the dog's reach. The trainer steps back about 6 ft. from the dog. After a pause of 10 seconds and when the dog's attention is on him, the trainer says the word OUT and begins a rapid, though not a startle-producing movement toward the food pan. The trainer picks up the pan and sets it down within eating reach of the dog. He then steps back a step or two and waits about 5 seconds. Now the trainer says OKAY and moves toward the dog, raises its head from the food pan by pulling at the animal's collar, and with the other hand, removes the pan from the animal's reach. The same sequence is repeated 5 more times. On the sixth OUT---food pairing the dog is allowed to eat all the food remaining in the pan. No special procedure is used to condition the verbal OKAY. Some trainers like to use it when they are ready to begin a trial or unit of work. It generally alerts the dog that rest, play, or eating time is over.

After about 2 sessions of pairing OUT and food, the dog may be observed suddenly to make anticipatory movements toward the food pan even before the trainer comes forward to give the dog the pan. Even if such signs of conditioning are not seen, learning has undoubtedly occurred. After two sessions, the verbal OUT can be used to acquire behavior and to maintain it in strength.

Step 3. Pairing OUT with Ball Play. When several varied pleasurable events are associated with the command OUT, it becomes a generalized reinforcer. In the present work, three different primary stimuli were available after the word OUT: a bit of beef and gravy in a pan, ball play, and a stroll through the woods. Since each stimulus was associated with a different location, the dogs often communicated what they desired for a payoff following OUT by moving toward a particular place.

In the procedure in which the command OUT is paired with ball play, the dog first is allowed to sample the ball reinforcer to determine if ball play is an attractive activity. The trainer can then throw the ball a short distance for the dog to chase. If the dog does not go after the ball, its attention and interest can be attracted by bouncing and rolling the ball in front of it. Verbal encouragement also helps. When the dog shows some enthusiasm for ball chase, then the pairing trials are begun. The trainer keeps the ball in the same pocket every time. With the dog close by, the trainer says OUT, reaches into his pocket, takes out the ball, and throws it in such a way that is bounces more than it rolls. The rewarding event brought on by the verbal OUT consists of 2 or 3 throws and retrievals. Not all dogs return the ball to the trainer, but go off with it and frequently lie down and chew it. For them, that form of play is uninterrupted for about 60 seconds. The trainer then goes out and takes the ball from the dog. Any attempt to force the dog to return the ball immediately reduces the pleasure of ball play. A total of 6 pairing trials in one session completes the conditioning procedure. The conditioned reinforcer OUT works in the same way as the conditioned reinforcer GOOD--it strengthens and maintains behavior by appearing at the moment of the behavior. In some circumstances, the trainer may elect to deliver more than one primary stimulus following the command OUT. For example, a dip in a pool might immediately follow ball play. Other pleasurable primary stimuli can be paired with OUT in the manner described for ball play and food.

<u>Conditioning the Aversive Stimulus TIME</u>.

The aversive spoken stimulus TIME is used both to punish undesirable responses and as a negative reinforcer to appear when a response is not forthcoming. The stimulus has no special conditioning procedure. The dog learns the significance of the signal during the shaping of behaviors. It follows badly executed responses, responses that are slow in coming or do not come at all, or when the dog emits an undesirable response. TIME is a generalized stimulus because several different events can follow it. For a behavior fault made during the performance of a response sequence which ends with primary reinforcement, the TIME signal means "begin the chained sequence over again." The most frequent event that follows TIME is "stand in place for the time-out duration." A time-out period of 30 seconds is effective. TIME can be followed by removal to a stake-out place where the dog is left alone for at least 10 minutes. Staking out the dog should not be done more often than 3 times a week. Under unusual conditions, as a TIME consequence, the dog may be immediately returned to its kennel and left there until the next test day. Under no circumstances is the time-out procedure made an excuse for rough or harsh treatment of a dog; the trainer must maintain a calm demeanor and deliberate manner in this, as in all other training procedures.

III. CONDITIONING ACCESSORY BEHAVIORS

Controlled Down Response

This training brings the DOWN behavior under control of a pulsed tone of 800 Hz A shaping procedure for obtaining the DOWN response is unnecessary because it is already present in the dog's behavior repertoire in the desired topography. Since the probability of the dog going down before training, however, is often very low, a procedure is used which forces the response in the early learning stages. Although, by definition, the forcing stimulus is aversive, it apparently does not elicit anxiety and fear emotions in the dog.

The DOWN training program includes control by the commands DOWN, STAY, and RISE. The program is so arranged that the dog learns in small easy-to-learn steps.

Step 1. <u>Adaptation to Training Setting</u>. A small training enclosure is selected. The trainer sits or reclines on the floor of the enclosure while the dog is free to move about off-leash. In a large enclosure, the dog is put on a short 6-ft leash. If the dog shows distress in the new setting, it is petted and verbally assured; otherwise, it is left alone to explore. Any spontaneous lying down behavior during these periods is reinforced. One session at this step is usually adequate. Thereafter, before a training session begins, the dog is permitted time to get adjusted to the setting while the trainer leisurely prepares to start the training trials.

Step 2. <u>Escape from an Aversive Stimulus</u>. The down response is forced by a procedure defined as "escape," i.e., the response removes an aversive stimulus. The training setting is arranged so that the only response which removes discomfort to the animal is dropping down into a reclining attitude. The aversive stimulus is a steady light pull downward on the dog's neck. The trainer may pet and verbally assure the dog during the early training trials.

A 4-ft nylon cord is attached to the dog's collar. The free end is put through an eye hook attached to the floor. Next, the collar is grasped under the dog's neck with the right hand and pulled gently downward while slack is taken up on the cord with left hand. The head is lowered only 3 or 4 inches from the level at which it is normally maintained. In this position, slack is usually found in the line, but is not taken up any more. In that predicament, the animal will be uncomfortable when standing but it will be relieved of discomfort when it lowers its body. When the dog starts to go down, the rope is released so that the dog can go all the way with ease. At the moment the forelegs touch the floor, the verbal GOOD is given followed by a bit of food. The hand with the food is held just out of reach of the reclining dog so that it has to get up to obtain the food. It is then ready for the next trial. The interval between a response and the start of the next trial can vary between three and five seconds. Individual trials last one minute if no response occurs. At the end of the minute, the dog is released from restraint for a period of about 15 seconds. The procedure is then repeated. At the trainer's option a trial can last ten minutes in the absence of a down response, and the animal is removed from the training setting. Thereafter, sessions end after 12 trials or ten minutes regardless of the number of trials

As learning progresses, the dog may begin to go down during the intertrial interval in the absence of the signalling stimulus. This problem can be corrected by immediately lifting the animal back on its feet. The dog will usually rise when a light upward pressure is applied on its abdomen with both hands.

Step 3. Learning to Anticipate the Aversive Stimulus. At this step, the dog learns to avoid the aversive stimulus by responding to a conditioned pre-aversive stimulus. The new stimulus is a light tug downward on the collar. It precedes the aversive, steady downward pressure by about 1/2 second in the manner of classical conditioning. Learning progress is tested on about every fifth trial by delaying the onset of the second (aversive) stimulus about 3 seconds. If the dog begins to go down during the delay interval, the aversive stimulus is not delivered. When some avoidance learning is demonstrated, the pairing procedure is continued by gradually increasing the interstimulus interval, allowing the animal eventually to avoid the aversive stimulus on every trial. The procedure of reinforcing the down response at the time the forelegs touch the floor is continued during the trials.

For this training, the 4-ft cord is replaced by one of 6-inch length. The procedures are conducted by the trainer from a kneeling position. The interval between trials can vary between 3 and 15 seconds. About 15 trials per session are adequate. Any unauthorized down responses are corrected by raising the animal back to the standing position. If at any time the pre-aversive stimulus loses its controlling ability over the down response, the pairing procedure is reinstituted.

Step 4. DOWN-STAY Training. The dog learns to stay in the down position for a pre-determined length of time before the verbal GOOD and food are delivered. At the moment the dog's front legs touch the ground, the trainer says STAY. The stay-in-place duration is gradually increased from an initial period of about 1 second. If the animal rises before the reinforcing stimuli appear, the down signal is repeated. Otherwise, the same procedures are used as in Step 3.

Step 5. Pairing a Tone with Tug-on-Leash. A new stimulus, a tone, will come to control DOWN behavior if it is paired with a stimulus which is already controlling the response. Since the on-time of the generated tone can be controlled, the pairing procedure for its conditioning is slightly modified. First, the tone is turned on, as required in the learning of any new stimulus. Then, 1/2 second later, the trainer begins the movement to grasp the short cord on the collar, followed by giving it a single tug. The tone stays on until it momentarily overlaps the tug signal. The response is reinforced either immediately or after a delay in the STAY position as in previous steps. Every fifth training trial is a test of conditioning, when the tug signal is delayed for about 3 seconds. As control by tone begins to be established, the interstimulus interval is gradually increased from 1/2 second to 3 seconds. Thereafter, the tone will be turned off at the moment the forelegs touch the ground. In this manner, tone cessation will control the stay response as does the voiced STAY. If the dog does not begin to go down at the end of 3 seconds, the trial terminates and a delay of 30 seconds is made before another trial is given. If the down response is made during the interstimulus interval the tug signal is not given or that trial. The pairing procedure is continued until the dog is responding to the tone signal alone. The trials are delivered with the trainer in the kneeling or sitting position.

Additional stimuli to control the down response can be conditioned at the discretion of the trainer by pairing stimuli in the manner described. The verbal DOWN, for instance, can now be paired with the tone or the collar tug-- the latter probably has better controlling properties-- which would hasten the learning of the new stimulus.

Step 6. Change in Trainer Stance. Learning by the dog to go down on signal, appears to be facilitated when the trainer works in a near recumbent position. In Step 6, the trainer assumes an upright stance. He may thus alter the work setting enough so that stimulus control is lost for a brief period. Control is easily re-established with patience and by following the procedures of Steps 3 and 4 with the trainer in the standing position.

Step 7. Rise Training. On occasion, an operational dog may be required to rise from the down attitude and continue an assignment. Although a specific signal is not used in the terminal system to raise the dog from the prone position, a signal can serve as a training aid for obtaining system behaviors subsequent to down-stay, such as recall, or a continuation of outward movement.

Conditioning of the verbal command RISE is made by pairing it with a stimulus which already controls the rising response from the prone position in most dogs, namely, a light tug upward on a leash. For those dogs that do not respond to such a tug signal, the pairing of the tug stimulus with the reinforcer, GOOD, will quickly correct the deficiency. An interstimulus interval of 1/2 second is maintained during all pairing trials. The reinforcers, GOOD and food, follow the rise-to-stand response. Any unsignalled down responses are corrected by immediately lifting the dog back on its feet, as described in Step 2.

Step 8. Conditioning RISE-STAY. The procedures of Step 6 are continued except that the command, STAY, follows the response instead of GOOD and food reinforcement. If the dog rises on command and stays in the standing position for a pre-determined time period, GOOD and food are then given. The required stay period is gradually increased over the trials, beginning with 1 second. For unauthorized rise-to-stand, the down signal is repeated.

Step 9. Transfer to Other Environmental Settings. When excellent control is established by the tone signal in one training setting, trials are continued in other places, especially in places where search chain trials are held. The down response will later be integrated into search chain trials. Occasional test sessions are conducted at the original training site until transfer to other settings is complete. The chaining of the down response with other behaviors is described in the section on search chain training.

Controlled Recall Response

The stimulus, GOOD, which is used to reinforce constructive behaviors by following them closely in time, also has a recall controlling property. This response occurs as part of a broader pre-consummatory behavior because its controlling stimulus (GOOD) regularly is followed by the appearance of food. This property of GOOD aids in the learning of signals more suitable for the particular purpose of controlling recall. The verbal command COME and a tone signal are learned.

Step 1. Pairing the Voiced COME and GOOD Stimuli. Recall training is conveniently conducted in a large room or enclosure, where the dog can roam some moderate distance from the trainer. The stimulus pair is delivered--COME is followed 1/2 second later by GOOD--when the dog's attention is away from the trainer. The trainer has a bit of food ready before the animal reaches him. On about every fifth trial the COME is given alone. If the return response is made on that test trial, the trainer delivers the reinforcer GOOD when the dog is about 6 paces from him. If no response occurs, the trainer readies himself for the next stimulus-pairing trial. The interval between the end of one trial and the start of another varies between 5 and 30 seconds. A session consists of 12 to 15 pairing trials. Additional trials can be given during play periods and while going from one work station to another.

Step 2. Discrimination Training to the Word COME. The procedure of Step 1 creates a generalization effect--recall is controlled by almost any verbal stimulus. For example, the dog will probably come to the trainer when he says FISH or any other word as well as when he says COME. A discrimination procedure easily corrects this effect. Simply, when the word COME is given, a return response is reinforced by GOOD and food; return following the presentation of other words, for example, BIRD or SKY, is not reinforced-- the response is ignored by the trainer. In this manner, the return response will eventually be made only to the command COME.

Step 3. Pairing a Tone Signal and the Verbal Command COME. The procedure is the same as that described for conditioning the verbal COME in Step 1, except for the on-time duration of the new signal. The tone comes on first, followed 2 seconds later by the word COME. The tone is turned off after the COME is sounded. Conditioning a new signal is still effective with a slightly longer interstimulus interval in the pairing procedure, if the first stimulus overlaps the second. The longer interval provides a test of conditioning on every trial. If the dog is observed to have made a commitment to return to the trainer within the 2 second interval, the verbal COME is not delivered. When learning is evident, trials are given with the tone command alone. The signal stays on for 3 seconds. If no return movement starts within this period, the trial terminates, and another trial is begun about 30 seconds later. Only return responses in the presence of the tone are reinforced. During this training, the return response may be elicited by other, spurious sounds, as is the case with many signals. Discrimination training to the specific character of the recall tone as distinguished from the DOWN and CHANGE DIRECTION tones is conducted after the latter signals are learned.

IV. DETECTION INDICATOR RESPONSES

Controlled Halt-Stand Stay

"HALT; STAND-STAY" was selected as a conditioned indicator response to the presence of people because it can be translated into a radio signal of clear meaning.

As a practical training matter, the detection response can be considered as the behavior sequence, "halt" and "stand-stay". A modified search chain, for which a restraining leash provides the differentiating controlling stimulus, is the vehicle used for this training. In this version of the search chain, the dog and the trainer move out together on the command MOVE, while in the normal search chain the dog moves away from the trainer.

Since the dog is under leash restraint it will usually follow the leash lead if the pulls and tugs on the leash are only slightly aversive. Thus, if a stimulus which means halt consistently precedes the restraining pull-on-the-neck stimulus, the dog will soon learn to anticipate the leash pull and will be able to avoid it by stopping in time. A collar is best for this work. If a choke chain is used, the leash is attached to the dead ring, which prevents the choke effect, and the trainer applies restraint less forcibly.

Learning can be brought about entirely by the escape and avoidance procedure, but the dog will receive ball play as payoff for a certain amount of work output. The conditioned reinforcer which precedes ball play is the voiced OUT.

Step 1. Conditioning the Halt Command, HUT, and STAY. Incidental to detection indicator response training, the dog learns to come to heel on the command HEEL. The trial procedure follows: While the trainer stands in place, he tells the dog HEEL. He follows this by leading the dog with a light leash pull to the heel position. At the same time, the trainer takes up and bunches the leash in his left hand, leaving no leash slack when the dog stands at heel. After a wait of 3 to 5 seconds, the signal MOVE is given, followed by a gentle tug forward on the leash. The trainer then leads off with his left foot. The HUT, meaning halt, is sounded after a variable distance of 20 to 30 meters is covered. It comes on at the moment the left foot touches down. One-half second later, a sudden full restraint from further forward movement is applied on the leash as the trainer also comes to a halt. As soon as the dog stops, the verbal STAY is given. About 5 seconds later, the procedure is repeated, beginning with the move out signal.

After 3 to 5 trials, the dog earns the reinforcers OUT and ball play. The OUT is sounded after the dog has been stand-stay for 3 to 5 seconds. After play, another series of trials can be conducted. Training continues by gradually eliminating the stimulus pairing trials until the halt response is completely under control of the verbal HUT.

Step 2. Training the Dog to Stand-Stay Away from Trainer. The dog learns to maintain the stand-stay when the trainer moves away from its side. The trials are otherwise conducted in the same manner as in Step 1. After the dog is told to stay following the halt, the trainer momentarily gives the arm-and hand STAY signal (palm of left hand placed in front of the dog's head), walks about

6 paces ahead of the animal, turns around and stops. Now the trainer again gives the verbal STAY and arm-and-hand signal (this time the palm of the hand is moved forward from the chest a distance of about 12 inches), and finally moves back to the heel position alongside the dog. After 3 to 5 seconds at heel, the entire trial is repeated. The interval during which the trainer remains some distance away from dog is gradually increased to 10 seconds. If the dog attempts to move while under the command STAY, the trainer immediately gives the HUT...STAY signal. The trainer next trains the dog to stand-stay, facing forward, after he moves several paces back of the dog. The distance is increased as performance improves. On about half of the trials, the dog is recalled and brought to heel where it awaits another trial. In the remaining trials, the trainer returns to the dog and continues from there.

Step 3. Transfer of Detection Indicator Response to People in View. Training in this step is conducted on a trail about 100 meters long. Every ten meters, just off-trail, a wood screen is placed behind which a person can hide. New elements introduced in this step are: (1) the pace of movement is a walk or a moderate trot, and (2) the "halt; stand-stay" is immediately preceded by the new stimulus, "person-in-view." Responses are reinforced with the verbal GOOD and food.

The trial is conducted as follows: An assistant is positioned behind one of the screens beside the trail, hidden from the dog's view. The trainer moves the dog out from heel at the start of the trail. When the assistant sees that the dog has advanced to within 10 or 15 feet of his position, he steps quickly out into the middle of the trail, immediately says HUT, then stands quietly in the trail without moving. One-half second after the HUT signal the trainer exerts a restraining pull on the leash to stop the dog. When the dog stops the assistant commands STAY; after the dog has remained in place for 3 to 5 seconds, the assistant says GOOD and comes forward to give the dog a pellet of food - the dog is permitted to break from the stand-stay position to receive the food. After the dog has been given the food, the trainer leads it back along the trail about 10 paces toward the starting point and keeps the dog facing that direction while the assistant now runs to hide behind another screen. When the assistant is set in place, the trainer turns around and commands the dog to heel. The trial procedure is then repeated. From 1 to 6 such trials may be conducted on a run along the trail; the assistant does not hide behind each screen on any single run. One or two runs comprise a training session.

In the early stages of personnel detection training, the assistant should be a familiar, friendly figure to the dog so that he does not elicit interfering avoidance behaviors. Trainers can serve as assistants for each other - they can play the role well since they are acquainted with the training procedures. This step is completed when the dog's halt response is evoked only by a person who suddenly comes into view of the animal. First the leash restraint and then the HUT signal are faded out gradually until the dog is no longer receiving either of these signals. The trainer does, however, continue to use the command STAY throughout the training program because the stand-stay-in-place response is difficult for the dog to maintain to the required criterion.

Step 4. **Transfer of Detection Response to People in Concealment**. Trials are run on a trail about 300 meters long with the tall grass or light undergrowth on both sides. The growth along the trail should be sufficiently dense to provide concealment for the assistant at numerous places. In order that the proper reinforcement contingencies can be maintained during learning, it is essential that the heretofore hidden assistant appear in the dog's view precisely at a given moment. This can be fairly well assured if the wind direction is mainly across the trail. The dog may encounter the assistant's scent as close as 10 to 20 meters from where the assistant is hidden. The main paired stimuli in this step are human odor and visualization of a person. A second pairing can be made with human voice sound and visualization. An interstimulus interval of 1/2 to 1 second should be maintained as closely as possible. Correct timing of the critical stimuli depends on knowing when the dog enters the scent cone. Thus, every effort should be made to get the dog to make a natural response to both human odor and voice sounds in the brush at the moment of their reception. At first, the dog may respond more strongly to sounds than to odor. The pairing procedure between the scent of a person and his visualization is effective if the person times his appearances in the dog's field of view to coincide precisely with the moment the dog manifests a natural response ("alert") to the scent. When this learning is complete, the dog will frequently show a brief natural alert response to human presence before emitting the conditioned halt and stand-stay response. The importance of an emitted learned detection indicator response to human presence is appreciated on those occasions when a weak natural alert is missed by the dog's handler.

In Step 4 trials, the dog is held on a leash by the trainer. The dog, however, is not kept at the heel position but is allowed to proceed ahead of the trainer by the length of the leash. This lessens the possibility that alerting responses will be suppressed in some way by the close presence of the trainer. The dog can be induced to proceed ahead of the trainer in several blank runs along the trail. If necessary, repeated trials from the heel position to a food bowl a short distance away can be conducted.

When the dog emits a natural alert on a human target trial, the assistant shows himself quickly but does not approach the dog. He follows his appearance with the command HUT. The trainer quickly restrains the dog from moving forward, and the assistant immediately commands the dog to stay. Following a hold in stand-stay position for 3 to 5 seconds, the trainer moves between the dog and the assistant, waits a moment there, then says GOOD, and follows this with a bit of food. While the dog is eating, the assistant disappears from view. The trainer then leads the dog out of the scent cone and begins the next trial down the trail from heel. Three human target trials per run are adequate. A second run on another programmed trail is desirable. The assistant does not move to another ambush position because it complicates the entire scent field—he may wait in the same position for the arrival of another dog. Step 4 training is completed when the command HUT and the restraining hold on the leash are faded out of the procedure and the assistant no longer shows himself—in this order of fade-out. Eventually, instead of the trainer interposing himself between the dog and the assistant on every trial before reinforcement is made, the trainer sometimes recalls the dog from the conditioned response position and reinforces it following its return. Finally, runs are made without regard to the wind direction.

Further guidance in the training and preparation of scout dogs for tactical employment is found in the US Army Field Manual 7-40.[4]

Controlled Sit Response

The sit response by off-leash scout dogs is the detection indicator of casualty producing devices, such as tripwire-activated bombs or grenades, buried mines, punji pits, etc.

The training outlined in this section obtains control of the sit response by a variety of stimuli. Response control by these stimuli is acquired by a classical conditioning procedure, i.e., new stimuli are learned by pairing them with stimuli which already control the response. Since the sit response is not under the control of any particular stimulus before training begins, the response is first forced by an aversive stimulus. In the aversive procedure (escape training), the dog learns to remove an irritant stimulus by sitting. Then, by the introduction of a pre-aversive stimulus, the dog learns to anticipate the aversive stimulus and to avoid it by making the sit response. No aversive stimuli are associated with the pairing of the remaining stimuli to be conditioned.

Step 1. Forcing the Sit Response. The most efficient positioning of the dog for sit training is just in front of the trainer where the trainer has easy control of the dog's head and forequarters with his right hand and its hindquarters with his left hand. Before the forcing trials begin, a period of adjustment is provided with dog in the training position. During this time any anxiety shown by the dog is reduced by verbal assurances and petting. For the sit training trial, the trainer brings the dog in front of him, takes up all the slack in the leash and bunches it in his right hand, places his left palm on the dog's rump and applies about 1 pound of pressure until the dog sits. The trainer waits 1 to 5 seconds after the dog is positioned before applying pressure. If the dog sits before the pressure stimulus is applied, it can easily be induced to stand if the trainer simply walks to another position a few feet away. The dog will usually get up and follow the trainer. Some dogs from the beginning respond almost reflexively to the pressure stimulus. Others may resist the pressure for long periods of time. Reinforcing with the verbal GOOD is accomplished when the dog's hindquarters touch the ground. Food is then held out beyond the dog's reach so that it has to get up to obtain it. The trainer has the option of reinforcing fractional responses, i.e., at any level of downward movement, but successively working toward the more complete response. A session usually lasts about 10 minutes. The trainer may give the dog a 15 second rest period after every minute of trial during which the dog does not respond.

Step 2. Pairing a Touch Stimulus with the Pressure Stimulus. The dog learns to sit whenever a light touch is applied to its rump. In a stimulus pairing procedure, the touch precedes the pressure stimulus by 1/2 second. Every fifth

---

[4] Scout Dog Training and Employment, Department of the Army Field Manual FM 7-40, March 1973.

trial tests for conditioning to the new stimulus by delaying the old stimulus be 2 seconds. If the sit response is initiated before the two seconds are up, the second stimulus is not delivered. When it is evident that the dog has learned to respond properly to the new stimulus, stimulus pairing trials are gradually eliminated. Usually 12 to 15 trials per session insure an acceptable rate of learning.

Step 3. Pairing the Verbal SIT with the Touch Stimulus. Voice control of the sit response is useful because it can be employed at a distance. The stimulus pairing procedure is the same as in Step 2--the new stimulus (the command SIT) precedes the old (touch on rump) by 1/2 second. When it is evident that the new stimulus alone evokes the desired response, the old one is gradually eliminated.

Step 4. The Transfer Stimulus. The scout dog (tactical dog) is called upon to locate a variety of casualty producing objects in its operational searches. The usual procedure by which a dog learns to respond to these stimuli is for the trainer to lead the dog to the object and command it to sit there. Clearly the trainer attempts, by this association method, to pass control from the verbal stimulus, SIT, to the object of interest. The pairing of two stimuli, a necessary condition of learning, often, however, does not have the intended result because a verbal stimulus may draw the dog's attention to the trainer and away from the stimulus object. Another object, rather than the verbal command, SIT, would serve better as the stimulus to control the conditioned alerting response. Such an object, called a transfer stimulus, is a 6" x 6" sheet of lucite upon which a black checkerboard figure is superimposed on a white background (Figure 1).

The transfer stimulus card, besides aiding the transfer of control of the sit response to other stimuli, is a convenient object to draw the dog in outward movement, to shape extended excursions away from the trainer, and as an aid in teaching the dog how to get over or across terrain obstacles. In all of these situations, food reinforcement occurs when the dog reaches the transfer stimulus card and sits at it.

When a dog undergoes detection training to a particular stimulus object in a field space that abounds with other stimuli, the trainer is often unsure if the detection response was made to the stimulus of interest. This problem is overcome in training by having the dog point to the stimulus object as part of its detection response. The transfer stimulus card is effectively used to train this behavior. "Pointing-to-the-object" behavior as part of the detection indicator response is readily eliminated when there is no longer any question about the object being detected.

Step 5. Adaptation to the Work Station. The trainer sits on a low stool, box, or even on the floor. The transfer stimulus card is placed on the floor in front of the trainer. A reinforcement contigency is arranged between direct movement by the dog toward the work station, i.e., the trainer/transfer card, and the verbal stimulus GOOD. During this shaping procedure the interval between reinforcements is about 2 seconds. The desired response, in this case, "close approach to the work station" can be obtained with 5 to 12 reinforcements.

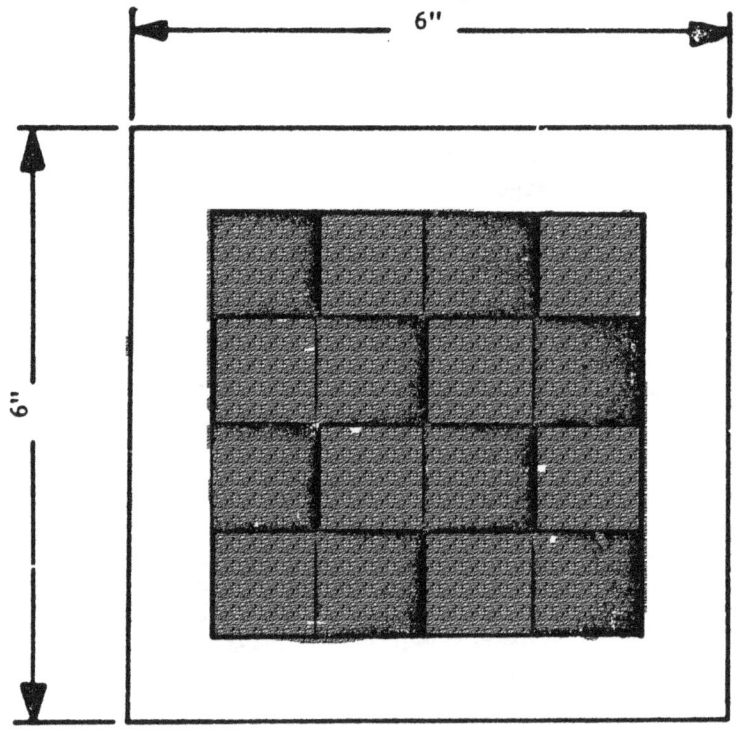

Figure 1. Transfer stimulus card

Step 6. **Shaping Head Lowering.** The shaping in this step goes through a series of responses which begins with a small movement of the head downward and ends with the nose touching the floor. The word GOOD is given at the moment the head moves the desired distance downward. The trainer selects progressively greater downward movements of the head for reinforcement until finally he will reinforce only those responses in which the dog touches the floor with its nose. Correct timing of the conditioned reinforcer, GOOD, is essential in this shaping procedure because response duration is extremely short. GOOD must always be voiced during a downward movement of the head or at the moment the nose touches the floor.

Step 7. **Shaping a Nose Touch Response to the Transfer Stimulus.** In Step 2, the animal may have occasionally touched the transfer stimulus card but it was not required to do so. Now, by differentially reinforcing responses, control is shifted from a nose touch anywhere on the floor to a nose touch only on the card. The dog is required to make a progressively closer response to the card position at a rapid training pace, though not so rapid as to lose the response. If at some point in the shaping procedure a sufficiently close response is not forthcoming the trainer may tentatively accept a more distant touch response for reinforcement.

Step 8. **The Transfer Stimulus Card Position is Varied.** This step will show if the dog is responding to a particular floor location with respect to the work station configuration or if it is responding to the transfer stimulus card. The card is moved away from its original position a short distance in various directions for a series of trials. After a few trials the dog should respond to the transfer card wherever it happens to be placed.

Step 9. **The Trainer Assumes a Standing Position.** In this step the trainer now stands, rather than sits at the work station, and he drops the card farther away from him than he would while sitting. If the change in position and the manner of dropping the card adversely affect the dog's newly conditioned behavior, the trainer must then make the transition more gradually. He can do this by changing his stance in small steps from sitting to standing and by gradually increasing the distance at which he drops the card. After a series of trials in which 2 or 3 reinforcements are made at each new position, full adaptation will occur to the new stimulus variable. At this point the trainer can reposition the card by hand-sailing it or by kicking it to variable distances at will.

Step 10. **Sit Response to the Transfer Card.** In this training step, the trainer works with two responses, each under the control of different discriminative stimuli. The dog learns to perform the responses in a chain sequence. The two stimuli, voiced SIT and "light-touch-on-rump," both of which control the sit response, are now used together as one compound stimulus to enhance the probability that the sit response will occur.

At the moment the dog's nose touches the card, the trainer gives the verbal SIT command, followed 1/2 second later by the touch-on-rump signal. When the hind quarters touch the floor, reinforcement is made with GOOD and food. Two or 3 nose touch responses are reinforced in succession with GOOD and food following each response while the dog remains in the sitting position. If the dog stands up following any reinforcement, it must then touch the card and sit again to obtain another reinforcement. By this procedure, the two responses soon blend into one--a single rocking motion forward to touch the card then backward to the sit position. When this happens, the sit response is under control of the transfer stimulus card and the other two sit-controlling stimuli are no longer used. Breakdown in the control of the response by the card is easily corrected by reverting to an earlier training procedure. When the nose-touch is no longer demanded as part of the sit response, it will naturally disappear from the dog's behavior topography.

Step 11. Pairing the Transfer Stimulus with Other Objects. After the sit response to the transfer stimulus card is incorporated into the search chain, as described in the following section, the stimulus card is then paired with casualty-producing devices following the classical conditioning procedure. This is a necessary and sufficient condition for obtaining control of the sit response by new stimuli. This section describes the conditioning of a ground cavity stimulus. It can serve as a general model for the conditioning of other stimulus objects.

A trail of about 100 meters long with 6 pits randomly situated along its length is prepared for the initial training. In the first series of trials, the transfer card is placed at the near edge of the pits in the trail. The dog is brought to heel at the start of the trail and moved out. The trainer does not follow the dog but remains in place. At the moment the dog sits at the card, the trainer says STAY and moves to the farther edge of the pit, turns, and faces the dog. One of two seconds later, he says GOOD and follows this with a morsel of food. The trainer then moves about five paces beyond the first pit and positions himself for the next trial. The dog should automatically follow and come to heel to await the MOVE signal. The next and remaining trials are run in the same manner as the first one. When the end of the trail is reached, trials are continued in the reverse direction

As learning progresses, the transfer card is gradually faded from sight. This procedure is begun by placing the card in the pit on some trials. It is continued in subsequent trials by covering an increasingly greater portion of the card with dirt. The dog is not corrected if it does not sit in the presence of a pit, but is permitted to continue along the trail. Transfer training is complete when the dog sits to the pit stimulus alone. Pit training continues on other prepared trails until the dog responds to a well camouflaged pit. During more advanced training, only one dog is run on any one prepared trail, to avoid the likelihood that dogs will respond, not to pits and other objects, but to places where other dogs sat.

V. THE SEARCH CHAIN

Basic search chain behavior conditioning begins with the command MOVE when the dog is standing at heel. Upon receiving the command MOVE, the dog proceeds to a food pan about 20 feet away, finds one or two bits of food in the pan, takes the food, and then returns to the heel position to await the next signal to move out.

The usefulness of the search chain becomes apparent when the behavior sequence is made to recur in a closed loop, i.e., following the consummatory response of one trial, the dog automatically comes to heel to begin another trial sequence. The cyclical ordering of the chain results in the rapid and efficient running of numerous trials. The food pan, a convenient vehicle for initially obtaining outgoing movement, is subsequently replaced by the transfer stimulus card, which is a more versatile training aid, and finally by hidden articles in realistic search scenarios.

Training is done in any indoor or outdoor enclosed area where the dog cannot leave the work setting. The smallest suitable area is an enclosure of about 250 sq ft (10 ft by 25 ft). Certain general guidelines should be observed: The dog is permitted to explore the enclosure briefly once or twice before any search chain training starts; the dog is also allowed a minute or two of free responding time in the enclosure at the start of a session; conditioning trials are begun at a performance level at least one step lower than the level reached at the end of the immediately preceding session; the session may end with a strong reinforcer such as ball play or a particularly palatable bit of food.

The training requires one deep food pan and a food dispenser. A commercial automatic feeder is the most convenient method of dispensing food pellets. The feeder is activated by the trainer from a hand held remote switch. If an automatic feeder is not available, the pellets can be dropped into the food pan by an assistant who stands back of the pan. The manual method requires a low amplitude noise-maker, such as a battery-operated buzzer, operated by the trainer by means of a switch. Noise from the feeder motor or from the noise-maker directly reinforces the responses selected for strengthening at every shaping step. The trainer times his operation of the feeder or noise-maker switch to coincide exactly with the dog's response. In the manual feeding method, the trainer's assistant drops a food pellet into the pan when he hears the sound come on. Precise timing of the food drop is not required.

During the learning of search chain behaviors, the trainer stands facing the food dispenser and pan which are about 20 feet away. He also stays at least 5 feet from any wall or fence so that the dog can circle around him.

Step 1. Stimulus Conditioning. The feeder motor noise or noise-maker come to reinforce behavior through frequent association with food. In the conditioning procedure, a sequence of several stimuli is initiated when the trainer closes the switch that activates either an automatic feeder or a simple buzzer or other noise-maker. In the case of the feeder, switch closure turns on the motor sound, which is followed shortly by a food drop sound, "PLUNK," and ends with a food pellet in sight. In the manual method the sequence is the same except that a food pellet is dropped into the pan by the assistant when he hears the buzzer sound. The food drop sound alone can be used to reinforce behavior in the manual method, but it cannot be precisely timed with the appearance of the dog's response--there is a time lag while the pellet drops.

Trials are conducted in series of 6. The interval between successive series is about 30 seconds and the dog must not be at the food pan when a new series begins. After the first trial of a series is given, the remaining 5 trials are spaced at 3-second intervals after the dog reaches the food pan. By the end of the first series, the dog should be responding to the food drop sound-- it immediately approaches the food pan at the sound. Conditioning to the motor noise (or buzzer) normally occurs after 2 or 3 trial series. The dog is then ready for behavior shaping

Step 2. Shaping Moving Toward the Trainer. The trainer, at first turns on the reinforcing sound every time the dog moves away from the food pan in any direction. An increase in the frequency of the moving-away response is quickly obtained. The next response to be reinforced is movement by the dog only in the direction of the trainer. If the dog at any time anticipates the signal and turns toward the food pan, the sound is not turned on. As conditioning proceeds, the required distance of travel toward the trainer increases progressively until finally the dog reaches the trainer.

Step 3. Shaping Movement Around the Trainer. Reinforcement is first made contingent on a nose-touch response to the trainer's outstretched right hand. The trainer can speed up the procedure at first by moving his hand toward the dog's nose until a touch is made. If feeder or buzzer sound onset occurs at the exact moment of the touch, the response is very quickly learned. The trainer withdraws his outstretched hand a short distance on every trial, and the dog must follow it to make the nose-touch response. As the trials progress the trainer lengthens the distance he withdraws his hand until finally the hand sweeps around his back. As the dog follows the hand as it sweeps behind the trainer, the feeder sound or buzzer is turned on without requiring a nose-touch. The dog's momentum brings it around the trainer's left side and toward the food pan. Next, the hand movement is faded out and the trainer simply stands quietly with a minimum of random movement. At this point in the shaping procedure, the dog moves in a fluid motion from the pan to the handler, around behind the handler from right to left and back toward the pan. The behavior sequence is performed at a rate of about 6 excursions per minute. The trainer continues to sound the reinforcing stimulus each time the dog circles behind him.

Step 4. Stop-at-Heel Training. The trainer positions himself next to the left sidewall of the enclosure in such a way that the dog has to come between the wall and trainer's left side close to the trainer. The transition from the old to the new position is made over a small number of trials.

The trainer physically stops the dog. He does this by encircling the dog's head with both arms as it tries to pass on his left. The dog is blocked from further forward movement by the trainer's hands placed at the base of its neck. The trainer grasps the dog's collar with his left hand, holds it for an instant, and lets go; the sound stimulus comes on about 1/2 second later. When the dog is stopped, its front feet should be even with the trainer's left heel.

Step 5. Stay and Move-from-Heel Training. At first the dog strains forward during the brief restraint and upon release of the collar it moves out. The stay-at-heel response is learned by timing release of the collar with the first lessening of forward straining. A new release signal, a slight forward flick of the collar, is introduced when the dog becomes less aware of the collar release. As with most conditioning stimuli, the release signal has both reinforcing and discriminative properties--it can shape the desired heel response topography through a series of changing performance requirements of the response, and it also sets the stage for the emission of the next response.

The duration of stay-at-heel when correctly done is about 3 seconds. If incorrectly done, i.e., the dog stands too far from the trainer's side, or its body is out of line, the trainer waits about 5 seconds, then he physically places the dog in the proper position and waits another 3 seconds before releasing it. In the later stages of training, the feeder sound turns on when the dog is about 6 feet from the pan. The only consequence for breaking from stay-at-heel is the withholding of reinforcement; the trainer does not recall the dog back to heel--the dog returns to heel on its own to begin the next trial.

When there is no longer any need to grasp the dog's collar - the dog stops without it - the flick of the collar is changed to a light tap on the back of the head. The command MOVE is taught by pairing it with the head tap signal. MOVE will come to control the release from stay-at-heel if it precedes the prior signal by 1/2 second in a series of trials. At this stage the trainer no longer need channel the dog close to him, so he can again position himself away from the enclosure wall.

Step 6. Adding a Detection Response to the Search Chain. The dog receives food either for going to the feeder or for sitting at a transfer stimulus card. It receives more food, however, and receives it sooner if it makes an appropriate sit response. This training is a continuation of training Step 10, of the section on controlled sit response.

The transfer stimulus card is placed in the dog's path in a search chain. In early sessions of this step the dog is given some warm-up trials without the search chain. When the dog reaches the card in either the warm-up trials or trials in a search chain, it is required to touch the card with its nose and to sit next to it. At the moment of the sit, the trainer says STAY and approaches the dog; he then stands facing the dog for a second or two when he says GOOD and gives the dog a food pellet. Three nose touch responses are reinforced in succession if the dog remains in the sitting position. If the dog stands up following the first or second reinforcement, it must then touch the card and sit again to obtain another. At the trainer's option, he can reinforce the third nose touch response with the feeder signal. Reinforcement is withheld for any response made to the card immediately following a reinforcement at the feeder. He repositions the card after each trial.

Step 7. Transfer of Search Chain to Other Settings. Trials continue to be conducted in the search chain training enclosure and also in one or two other settings daily. The first six trials in the enclosure continue to be run as in Step 6. For three remaining trials, the trainer positions himself with his back to the feeder, so the dog is sent out in the reverse direction toward the stimulus card. Thus, the dog begins to learn to move straight out in the direction it is facing at heel.

After the last trial in the enclosure, the trainer moves to another near-by setting where runs of 30 to 50 meters can be made. He begins training there by placing the stimulus card within sight and gradually shaping longer distances to the card. Three trials are programed. The first two are reinforced with food pellets in the standard manner. On the third trial, the correct sit response is reinforced with OUT, followed by ball play or by a small amount of extra-palatable food in a bowl.

Immediately following the last reinforcement in the previous setting, the dog is taken to a road or trail where runs of a greater distance can be made. As before, the early trial runs are short and increase in travel distance is made with trial experience. The dog makes only one run in this greater distance setting. The reinforcers are OUT and several minutes of chewing on a rawhide bone, or any other attractive reinforcer. After two weeks of training in the three settings, the original one may be dropped from the schedule. When good performance is attained on a 100-meter length of road or trail, the search chain procedure is used to obtain outward movement, a change in direction of movement, and search and detection behavior in a variety of realistic settings.

VI. CONTROL OF OUTGOING BEHAVIOR IN A FREE FIELD

Remote control of a dog implies that the handler can obtain outward movement by the dog, changes in its direction of movement, and cause it to return to him, either by commands he gives or through the action of terrain stimuli which are partly arranged by him. The dog's locomotor responses while in a field space are for the most part continuously under the control of terrain stimuli--the handler's commands mainly serve to bring about a shift of control from one terrain feature to another.

In this section, procedures are described for obtaining continued outward movement, i.e., the dog generally stays on the course set when it moved out from the trainer's side. Roads and trails help keep the dog in bounds while long distance travel is shaped. The dog also comes to depend on distant environmental sitmuli, such as prominent features ahead in its visual field, to control the direction of its excursions across open fields.

Shaping Long Distance Runs.

A stimulus card may be placed at a given location and the trainer starts a search chain trail at increasingly greater distances from the card. This technique is apt to promote place learning and therefore it must be used with caution and with this in mind.

An alternative method that does not encourage place learning, makes use of a changing card position, while the starting location remains the same. Over successive trials the dog travels increasingly greater distances to the card. During the early shaping period, the distance change from one trial to another is made small--the dog should be able to see the card from where it was located on the previous trial. In three of four trials, the trainer goes to the dog and reinforces sitting-in-place, while in one trial of four, the dog is recalled followed by reinforcement. If the dog is recalled more frequently before reinforcement, it will begin to anticipate the signal and break from sit before the recall signal is given. It seems certain that realistic tactical doctrine would require a handler to recall his dog for reinforcement following a detection response. In training, and in maintenance exercises, however, reinforcement normally occurs at the response site (reinforcement-in-place). The efficiency of running moderate to long distance trials can be increased if an assistant is called upon to reinforce the dog in place. The assistant waits out of sight downwind of the stimulus card; he calls out STAY when the dog sits at the card and then reveals himself, comes forward and feeds the dog. Following reinforcement, the trainer recalls the dog, pets it and begins another trial. The first trial of a session is run over a shorter distance then that run in the last trial of the preceding session. The longer the distance travelled by the dog, the larger the food payoff.

In yet another variation of the method, both the trainer's position and the position of the card change on every trial. The card position of one trial becomes the starting point for the next. The trainer takes up his new position about 6 feet beyond the most recent card location. Six runs of equal road length are made in a single session, and the starting point for every session is the same. When the total distance in six successive runs reaches 1/2 mile; then the same 1/2 mile distance is covered in gradually fewer runs, until finally the distance is reached in one run. About 3 more sessions on the familiar stretch of road completes the training there. The same procedure is then followed on other long distance road segments.

Control of Movement in Open Fields.

The dog learns to move across an open field toward a prominent feature of the landscape. It is possible that a dog while standing at heel waiting to move out across an open field scans the field and selects a target object as a guiding landmark before it moves out.

In training the dog to move toward a treeline or a brushline, the demarcation between dense and scrub vegetation snould be distinct. In the short distance runs made during early training, 12 transfer stimulus cards are lined up along the field boundary. The distance separating them should be about equal to the length of the runs to be made toward each of them. The initial distance between the move-out position and a card is about 2 meters. At this distance the dog can probably see the card. The trainer makes sure that the dog's body is perpendicular to the woodline before the trial begins; he moves the dog out and, when it sits, commands STAY, walks to the dog, faces it fcr a second or two, says GOOD and presents a morsel of food to the dog. At this point the trainer picks up the card and then positions himself in front of another card and starts another trial. All excursions in any session are usually of equal length. Increments by which succeeding series of runs are increased can be progressively, 2, 4, 6, 10, 20, 30, 40, 50, 75, 100, 150 meters, etc. For the longer runs (50 meters and more) 5 stimulus cards are placed along the woodline about 20 meters apart. The dog is aimed at the 3rd card in the line. This procedure increases the probability that the dog will come across one of the cards at the moment it reaches the woodline. Whenever the dog tends to wander from a direct path to the woodline the trainer should shorten the distance to be traversed to reach a card. If the dog intercepts the woodline at a point where there is no stimulus card it is commanded to halt, the trainer walks quickly to it, stands facing the dog for a second or two, say GOOD followed by brief petting and praise then returns the dog to the starting point of the last trial or to one a little closer to the card, and repeats the trial. It should be noted that in this case - when the dog misses a card - it does not receive food. As the length of the runs is increased the number of trials in a session is decreased.

Lengthening the distance over which a dog must proceed toward a prominent object in an open field is shaped in a similar manner. If the object is fairly large, as a building, transfer stimulus cards are lined along the side that the dog will approach so that the dog will find a card without having to search. The object is approached from different directions on successive trial runs of gradually increasing length. It is convenient to use a transportable target panel about 2 ft by 2 ft mounted on a frame standing 4 ft high and painted in the same black and white checkerboard pattern as a transfer stimulus card as a target object in guidance training. When a target panel is used a small transfer stimulus card is placed next to it.

Learning to Overcome Obstacles.

Fences, brush, tunnels, waterways, etc., may occur in the path of a dog's excursion. Training in how to overcome obstacles in its path gives a dog a real advantage over one that has not had such training. Obstacle training is conveniently accomplished in a restricted setting. A long, narrow enclosure is arranged in which the dog must proceed from one end to the other at which a stimulus card is placed. Physical barriers are then positioned in the path of the dog. The barriers are made more difficult to traverse in succeeding trials For example, a fence can be increased from an initial height of 6 in or less to a height of 6 ft or more in small increments.

A waterway or stream presents a special problem in that the training must of necessity be done at the site of the water barrier. A road or trail partly under water is especially advantageous in this training because it provides directional guidance for the dog. Training a dog to cross a stream that runs through a field may take a little longer but the method in any case is essentially similar. A target panel and a transfer stimulus card are positioned about 50 meters from the edge of the water. Trials are begun from a point near the target panel between the target panel and the water. The distance the dog must travel to the target panel is increased incrementally as the trainer moves the starting point closer to the water. The starting point is brought to the near edge of the water, then part way into the water and finally trials are begun from the far side of the water, forcing the dog to cross the water to reach the target panel.

VII. REMOTE CONTROL

FIRST EXPERIMENT. EXPLORATORY SHORT METHOD

In this experiment the objective was to induce the dog to change direction during an excursion in a cross-shaped training structure in response to a sound signal. The training structure similar to Figure 2, consisted of 4 alleys, each 25 ft long and 8 ft wide, arranged in a cross. Where the alleys joined was called the choice point. The outer end of one alley served as the starting point for all trials. The inner end of each of the other 3 alleys, at the central choice point, was equipped with a gate that could be opened and closed at will to form several different maze patterns. A black-painted box, measuring 18 in. in all dimensions, mounted 24 in. above the ground, and containing a 150-watt globular, fronted light bulb that operated in a flashing mode, was placed at the outer end of each of these legs. A food pan was positioned beneath each box. An amplifier-loudspeaker was installed at the starting position to sound the tone signal. Hand-held switches controlled the tone signal and the lights. Four dogs were used in this experiment. It was reasoned that a dog would have little difficulty in first learning to guide on a flashing light at the end of an alley where food was present. Then, it was thought that when the light stimulus was faded out of sight, guidance control could be passed to a tone signal that would come on whenever the dog turned into an alley other than the one in which food was present.

Step 1. Adaptation to the Cross. Adaptation to the cross consisted of permitting the dog to investigate each alley with the trainer nearby. When the dog moved freely through the alleys and showed no avoidance responses, adaptation continued with the programming of food pellets at the end of a single open alley. Trials were begun from the heel position in the starting alley. On the command, MOVE, the dog moved along the alley to the central choice point from where it could proceed only into the alley with an open gate. At the outer end of the open alley, the dog found food pellets in the food pan, placed there by an assistant before the trail began. The dogs were run on all 3 alleys in the single-alley configuration in each of several sessions. When the dog's excursions had become smooth and deliberate with little hesitation, the dog was adapted to the flashing light. The trainer simply dropped food into the pan or hand-fed the dog at the intelligence panel with the flashing light on. The next step was undertaken when the dog showed no avoidance to the light.

Step 2. Two-Position Discrimination. Three alleys provided three ways in which two alleys could be combined: right-center; left-center; right-left. Only one 2-alley combination was used in a given session. Trials were conducted as follows: An assistant silently placed one or two food pellets in a food pan in one of the alleys while the trainer held the dog facing away from that alley. The dog was then brought to the starting position and moved out. The flashing light was turned on as the animal moved toward the choice point and remained on until the food was removed. A correction technique was used whereby the dog was permitted to move from the wrong alley into the correct one without first coming to heel. Each session consisted of 12 to 15 trials.

None of the four dogs learned the light discrimination. Position orientation became a common problem. With any two alleys, the dogs consistently entered

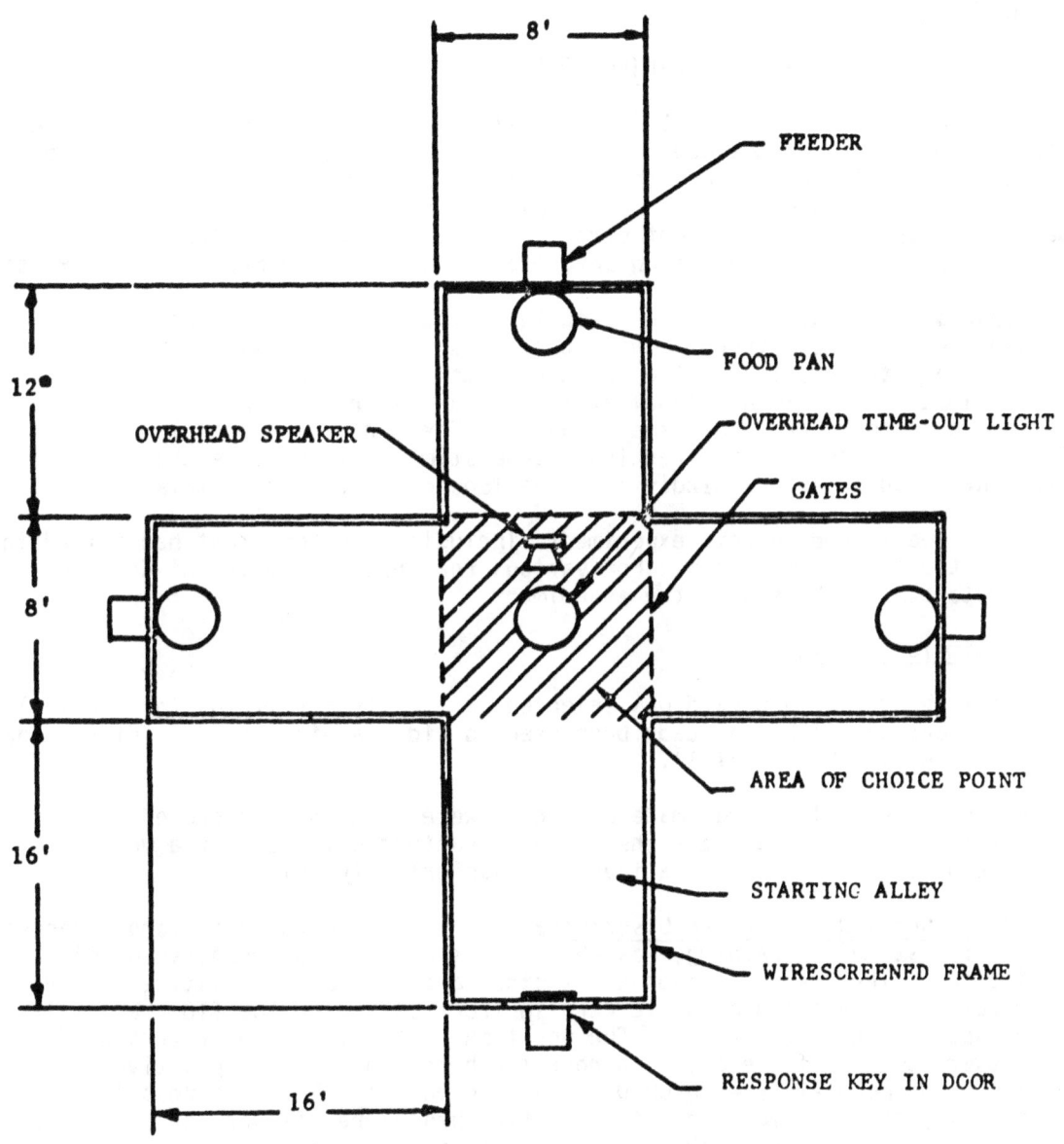

Figure 2. Cross-shaped apparatus

one alley in preference to the other. Only one of the four dogs continued to work steadily until the end of the experiment; the other dogs began to refuse to work during the sessions. It was concluded that behavior suppression occurred as a result of heightened anxiety generated in the animals by the repetition of errors. After six weeks of effort, the experiment was terminated and another experiment was designed.

SECOND EXPERIMENT. MANUAL TRAINING PROCEDURES.

It became evident that the successful development of complex, tone-controlled avoidance movements by the dog depends on techniques which produce learning virtually without error. Once this was recognized, an experiment was designed in which the shaping steps were made small and easy to learn, with a criterion goal of no more than 10 per cent error at any learning step. Thus, almost flawless performance was required before a change could be made to the next step

The first discriminative stimulus used to control change in direction was a painted checkerboard pattern. This acted as a directional signal to the dog by reason of its location in the training setting. By pairing this pattern with a lamp--a directional signal at first, but later made omnidirectional-- the lamp came to control the same behavior. The omnidirectional lamp was finally paired with an omnidirectional tone stimulus to produce the terminal tone-controlled behavior. Individual sessions consisted of 20 trials.

Eight dogs were used in this experiment, including the four that had failed to learn in the first experiment. For various reasons, the number of dogs was reduced to two by the end of the experiment.

Basic Stimulus Conditioning.

The checkerboard pattern card was selected as the first stimulus for controlling direction because it had already been used to aid learning of other discriminations and skills (Section III).

Two 6" square stimulus cards made of lucite were used in this procedure--one, the positive card, had a black checkerboard pattern painted over a white background; the other, the negative card, was entirely white.

Step 1. Conditioning Pattern Discrimination. The nose-touch-to-card, learned in transfer stimulus training, was chosen as the indicator response of discrimination. The trainer sat on a low stool and placed the positive checkerboard card and the plain negative card in front of him on the floor about 12 inches apart, center to center. The position of the cards was reversed after about every two reinforcements. A nose touch response to the positive card was followed immediately with GOOD and a food pellet. The same response to the negative card was ignored. A correction procedure allowed the animal to shift from the negative to the positive card so that every trial would end with reinforcement. Simultaneous presentation of the stimuli was occasionally changed to a successive presentation in which only one stimulus card was presented in a trial. By this alternate procedure, if no response occurred in five seconds in the presence of the negative stimulus, the trial ended and another presentation was made. When discrimination learning was unmistakeable, card discrimination trials were continued as a warm-up procedure for sessions of the following step.

Step 2. Transfer of Pattern Discrimination to Excursions in a Runway. This step was designed to produce differential excursions toward a checkerboard pattern and away from a plain one.

This work setting was a straight runway (Figure 3). The distance from the starting point to the food pans was about 20 feet. The pans were set 4 feet apart, center to center. Stimulus cards, each measuring 20 inches square, were placed upright back of the pans. In early trials, the 6-ft long barrier shown in Figure 3 was not used; it was assumed that the dogs would be guided by the stimulus cards from the beginning. As it turned out, however, the dogs initially responded on the basis of position, as in the first experiment. The barrier was then emplaced to increase the response cost to the dogs of going first to the negative stimulus. The trainer's assistant (programmer) stood just outside the runway during trials where he could conveniently reach in to change cards and place food pellets in the appropriate pan.

The assistant switched cards on a random basis after one or two trials, and placed food pellets silently in the appropriate pan between trials while the dog faced away from the program field. Each trial was run as follows: The dog moved out from heel, on command, to the stimulus position, took the food and immediately returned to heel for the next trial. In the event a dog made an incorrect discrimination and went first to the negative position, it was allowed to correct the mistake and go at once to the positive position without returning to heel. Most dogs required 12 sessions in which to learn to go to the positive position without first testing the negative side. They appeared to work with enthusiasm on the problem and none of the dogs showed evidence of quitting during a session.

Step 3. Learning Light-Dark Discrimination. A stimulus pair consisting of a non-illuminated light bulb (the positive condition) and a flashing light bulb (the negative condition) was introduced next as a directional control. The lamps were recessed within boxes in which the open face was 18 inches square. The interior of each box was painted dull black. The light boxes were positioned 20 inches from the floor at the end of the runway just above the stimulus cards. The trainer controlled the flashing light by means of a hand-held switch. With respect to learning the correct discrimination, it was irrelevant whether it occurred through avoidance of the flashing light or approach to the non-illuminated bulb. In any event, the flashing light was paired with the blank (negative) card and the non-illuminated bulb was paired with the patterned (positive) card.

The trial procedure was as follows: The trainer switched on the flashing light about 1 second before the dog was released from heel, and turned it off when the animal lowered its head into the positive food pan. For the purpose of adaptation, the light intensity was made progressively brighter over the course of the trials until it reached its full 100-watt intensity. Then, the checkerboard pattern, in a fading procedure, was made increasingly lighter over 8 shades of cards until it looked like the negative card with no pattern. Whenever the dog began making errors of discrimination, the next darker patterned card was reintroduced for a series of trials. At the end of this training step, the dogs were going consistently to the non-illuminated side without checking out the other side first. It was observed during the fading procedure that a dog would move its head back and forth as spectators do at a tennis match, just before being released from heel.

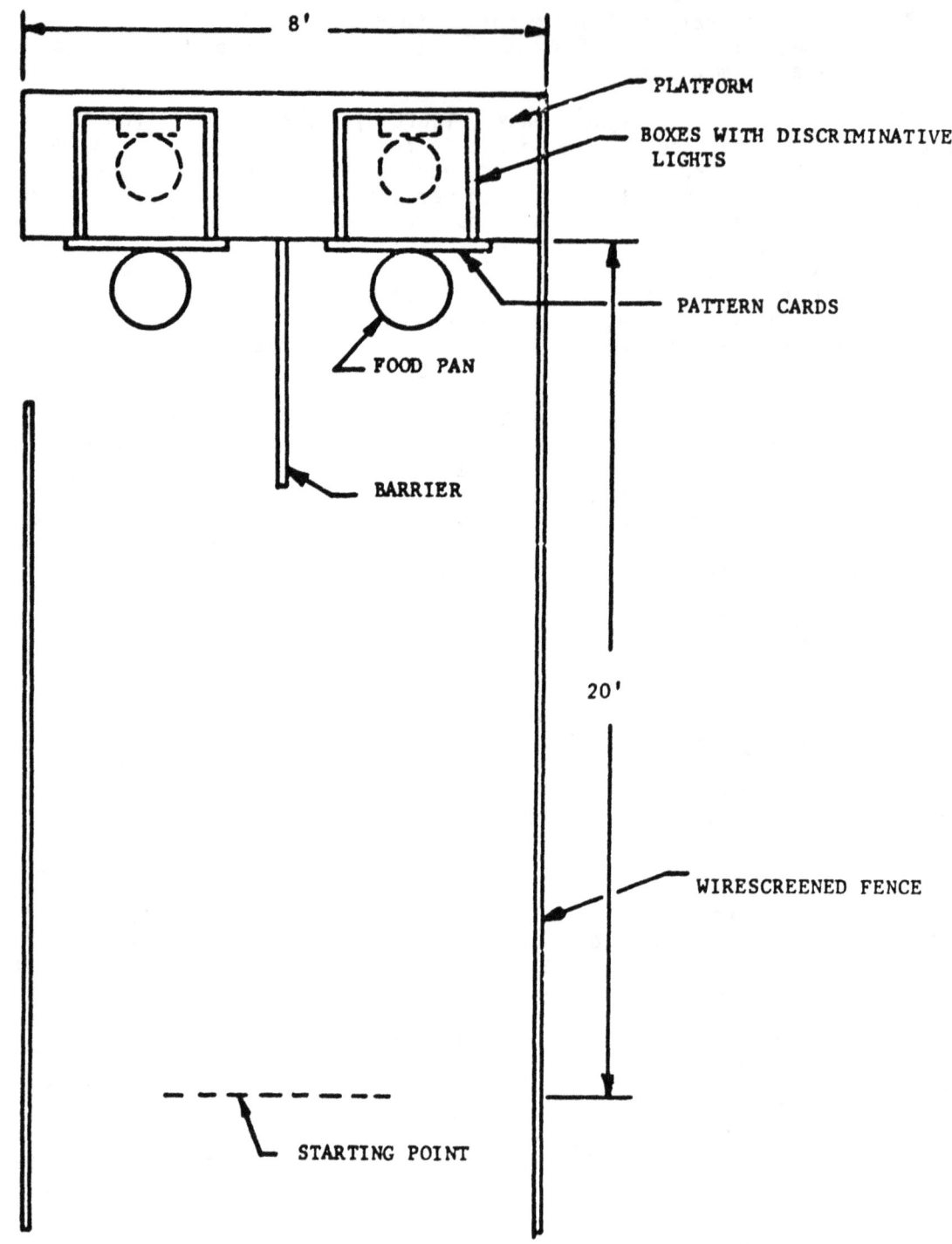

Figure 3. Straight alley apparatus

Step 4. Reducing Reaction Time to Light Onset. Two kinds of trials were run in this step. In non-discrimination trials the flashing light was not turned on so that either terminal position was correct. In these trials the assistant dropped one or two food pellets in the food pan that the dog elected to approach In discrimination trials, on the other hand, the light was turned on after the dog was released from heel but before it entered the choice point of the runway, designated as the area four feet in front of the barrier. The distance of travel before light onset was progressively increased. The objective of the non-discrimination trials was to obtain continued outward movement when no direction-controlling stimulus was present. The ratio of non-discrimination trials to discrimination trials was gradually reduced until the non-discrimination trials were finally eliminated.

Transfer of Training to a 2-Choice Construction.

In this step, the dog learned to escape the flashing light when it came on by moving in another direction. Training now was conducted in a Y-construction, shown in Figure 4. The starting runway or alley was 16 ft long and 8 ft wide. Each of the two arms, or the "choice" alleys, was 16 ft long and 4 ft wide. The choice point was taken as the area about 4 feet in front of the bifurcation. A food pan and a light fixture were emplaced at the end of each choice alley.

Step 5. Escape Conditioning to the Flashing Light. Dogs were adapted to the new apparatus in two steps. First, a series of trials was conducted in which the trainer turned on the discriminative light before the dog moved out from heel. Then, in a series of trials, the onset of the light was delayed until the dog was in outward movement, but before it reached the choice point. Non-discriminative trials, in which either alley was correct, as in Step 4, were also performed.

Until this step, the signal light had been turned on in every discrimination trial. Now, however, the light would be expected to come on in only 50 percent of the discrimination trials. The light was turned on only when both of two response conditions were met: the dog reached the choice point, and it showed a clear commitment to go into the incorrect alley. The light was turned off at the moment the dog started to change direction away from the incorrect alley. If the dog changed back to its original direction into the incorrect alley, the light immediately was turned on again.

Step 6. Introduction of an Omnidirectional Flashing Light. The discriminative flashing light that until now had appeared only at the end of a runway in the training construction, was replaced in this step, by an overhead flashing light that illuminated the entire training field. The purpose of this change was to give the flashing light stimulus the same omnidirectional character as the tone would have when it was paired later on with the light.

A 150-watt lamp fixture was centered above the choice point of the Y-construction. It was controlled by the same hand-held switch that controlled the directional alley light, so that both lights came on together at the press of a single button. Both lights were turned off when the dog changed direction to enter the correct alley.

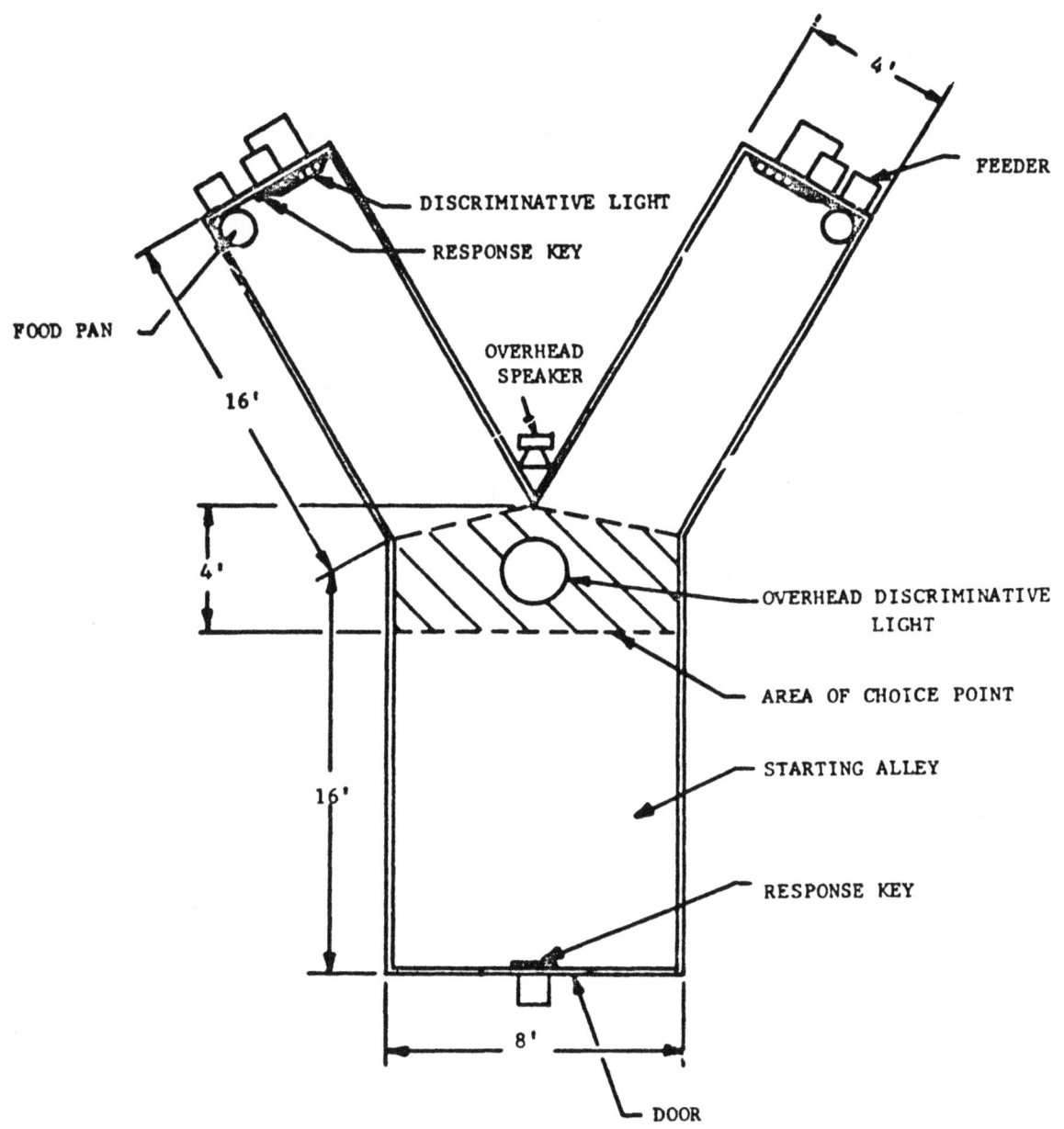

Figure 4. Y-shaped apparatus

Following one adaptation session with both lights at full brightness, the directional alley light was gradually faded out of sight over several sessions. When errors began to occur, i.e., when a dog persisted in entering the incorrect alley in the presence of the flashing light stimulus, the intensity was increased for a series of trials and then again faded.

No unusual problems arose during the transfer procedure. Six dogs learned to respond to the omnidirectional light in 10 to 20 sessions. Two dogs were dropped from the program at this time.

Step 7. Transfer of Direction Control from Light to Tone. The onset and cessation of the new stimulus, a tone, were made contingent on the same responses as was the light in the previous step, namely, movement toward the incorrect alley and change of direction away from the incorrect alley, respectively. Onset of the light stimulus, in this step, was delayed 1/2 second and was turned off at the same time as the tone. Gradually, the interval between the tone and the light was lengthened; not to test for learning, but to increase the response cost to the animal if it persisted in an excursion into the incorrect alley when the light came on. If a correct response was made, i.e., if the dog reversed its course and moved toward the correct alley, in the interval between stimuli, the light did not then come on. Learning in this step was rapid. Control by tone alone was established by the end of 6 training sessions.

Transfer of Training to a Cross.

The same cross-shaped construction was used as in the first experiment. The lamp fixtures at the end of each alley, however, were removed.

Step 8. Learning When to Stop Changing Direction. Adaptation to the setting included a period of exploration and running search chain trials in single, open alleys. Training then was conducted first on various 2-alley combinations right alley-center alley; left alley-center alley; right alley-left alley. A single 2-alley combination was used throughout a session. Finally, all three alleys were opened.

In the Y-construction used in the previous step, the signal was turned off only to reinforce the response of turning away from the incorrect direction. Thus, tone onset alone not only identified the incorrect direction but, indirectly, it also indicated the correct direction in which the dog was to proceed. In order to maximize informational properties of the tone signal, signal onset was made to mean "start turning" and signal cessation, "stop turning." This meaning of signal cessation was conditioned in the cross-construction where, at the choice point, the dog had 3 directional options.

Turning off the tone when the dog's body was in line with the correct alley was found not to work well. An analysis of the situation showed that the dog moved ahead in the direction its head was pointing at the time the signal was turned off. Learning was obtained when the observing response was selected for reinforcement by signal cessation.

In 10 of 20 trials (50%) that comprised one session, the center alley was the correct one. This schedule encouraged the dogs to continue outward movement without turning in the absence of a turn signal. If the dog returned to heel directly from an incorrect alley, the trial was repeated. Also, if the animal entered an incorrect alley, the signal was turned off when its head pointed toward the choice point, and was turned on again only if the dog entered another incorrect alley. Following learning, the cross construction was used periodically to maintain the turning skill.

THIRD EXPERIMENT. AUTOMATION OF TRAINING PROCEDURES.

The training program, as developed in the Second Experiment, though successful in obtaining controlled changes in direction by dogs, still was essentially experimental. It became clear that the various contingencies, or the relations between behavior and its consequences, could not be precisely arranged by manual training procedures that, by their nature are often imprecise.

The logical solution to this problem is to adopt a computer control system, such as is already being used by some workers in the behavioral field. Such a system, with unequalled precision, runs experiments that require the presentation of stimuli, the accumulation of response information, the issuance of control signals to reinforcement devices, and the command of appropriate print-out devices and recorders. Besides promoting rapid learning and uniformity of performance by dogs, automated training methods require less manpower and fewer skilled personnel, provide uniformity of procedures, and simplify the presentation of procedures that involve small incremental training steps.

In the initial phase of this experiment, automatic equipment was used in an attempt to obtain responses to turn signals by a classical conditioning procedure. It became apparent that the relatively simple procedure of this phase was not effective, however, in developing the desired behavior. Subsequently, other automated procedures were introduced in order to develop the necessary tone-controlled turning skills.

Conditioning Stimuli by Pavlovian Methods.

Since events in the animal's environment can be well controlled by logic operations, an attempt was made to simplify and automate some early training procedures. The experimental design called for the acquisition of behavior control by several stimuli within a Skinner box setting. The first learning was movement toward a feeder noise and the sound of a food pellet dropped into a pan. The strategy then was to obtain conditioning of movement toward a light. Finally, in a simple stimulus reversal procedure, movement would be obtained away from the light.

Training was conducted in an environment similar to the one shown in Figure 5. A lamp was mounted on a panel at each end of the box. Its molded translucent screen measured one foot square and extended 1 inch into the box. End panel keys, shown in the figure, were not used initially. A transilluminated response key, that measured 6" x 8", was mounted midway on one of the 8 ft walls 30 in above the floor. Four dogs, none of which had been used in the earlier training experiments, served as subjects in the present experiment.

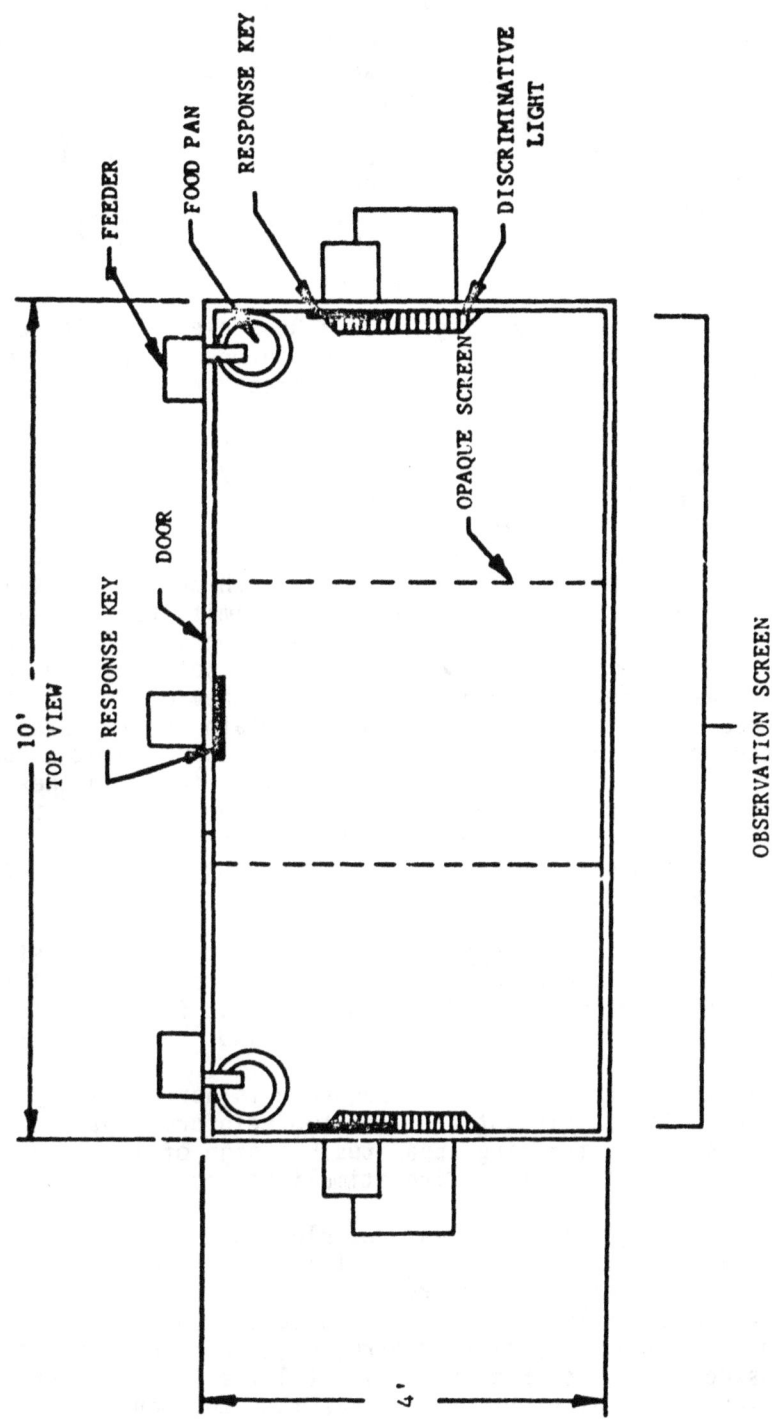

Figure 5. Skinner box environment

Step 1. Adaptation and Feeder Training. The dog was given an opportunity to explore the training box for at least one short session. The trainer provided a calming effect on the animal by accompanying the dog into the box or by staying close to the screened wall outside the box. He also hand fed the dog at this time. During feeder training, one or the other food dispenser dropped a pellet of food into a pan in a random sequence. An interval of about 5 seconds between food drops was controlled by a variable timer. Conditioning to the food drop sound occurred within about 6 presentations. By the end of 20 trials, all dogs were moving quickly in the direction of the feeder motor sound.

Step 2. Feeder Stimuli Made Contingent on a Response. The feeder motor noise and PLUNK sound of a pellet hitting the pan depended on a key press response in this step instead of on a clock. The key press started a stimulus sequence which ended with food-in-mouth.

The key response was shaped by first differentially reinforcing ever closer approaches toward the key. In the absence of a key response, to operate the feeder automatically, a hand-held switch was employed. A dog's characteristic habit of simply sniffing at the surface of an object, rather than "nosing" it, delayed the desired response shaping of "nose-touch-on-key," until slight forward movement of the head, in addition to proximity to the key, in sort of a pecking motion, was selected as the response topography for reinforcement. The key touch then became conditioned. Finally, displacing the key inward about 1/8 inch was achieved. Because a natural behavior interfered with the learning process, it took about 6 sessions to obtain the key press response. A change in the key design made at a later time brought about a substantial reduction in key shaping time. During the shaping steps and thereafter, a response operated one or the other feeders on a random basis. Subdued light illuminated the box interior. The key light remained on for the session duration. The session automatically terminated after 60 reinforcements. The trainer then removed the dog from the box.

Step 3. Light Discrimination Training. A light became the first stimulus in a stimulus sequence started by a lever press response. The light turned on at the location where food would be found; it was followed 1/2 second later by the feeder motor, next, by the PLUNK sound.

The response which starts the stimulus sequence is appropriately made at a key position which is equidistant from the two reinforcement locations at the opposite ends of the training apparatus. A sign of learning was movement in the direction of the discriminative stimuli before food appeared.

Conditioning to the light stimulus was not clear after 2500 trials. The animals typically hesitated at the key position until the motor sound came on. During the next 500 trials, the interval between the light and motor sound was increased to 2 seconds. This also failed to obtain conditioning to the light. The increased delay produced either control by a spurious contingency, under which the dogs continued to press the key until the feeder motor came on, or a position orientation which brought on immediate movement toward one pan following a single key press. After 3000 trials, this phase of the experiment was discontinued.

Apparently, the pairing of stimuli in the manner of classical conditioning is not a sufficient condition for dogs to learn a visual discrimination. The failure to obtain learning of a visual task suggests that the visual apparatus is not a primary sensory pathway for stimulus perception for a dog. It appears that a dog will not use its sight to perceive events that lead to a reinforcing effect if other stimuli, i.e., olfactory or auditory, are present and lead to the same result. This phenomenon is dramatically demonstrated by observing a dog search for food bits, accidentally spilled near its food dish. Although the particles are in plain view, they are picked up only after a random search when the dog's nose practically touches them. On the other hand, there is ample evidence to show that dogs can and do use visual stimuli effectively. For example, in a clear instance of Pavlovian conditioning, dogs quickly learn to anticipate the appearance of food when they see the trainer's hand come out of his food pouch. Begging is observed before the food appears. Conditioning of visual stimuli will also occur when their presence is critical to the performance of some rewarding work. Thus, a dog learns the meaning of visual stimuli which inform it when responses will be followed by a pleasurable consequence. This was shown in a situation where a nose touch to a patterned card was followed by food and the same response to a non-patterned card was not. In another case, visual stimuli which define a response-reinforced period, must be turned on by the animal. In other words, the animal turns on the visual indicator of a period when responses will be reinforced. In the next phase of the present experiment, dogs turned on a visual stimulus that not only indicated to them a period when work would be food-reinforced, but also indicated to which of two alternative positions the reinforcement condition applied.

Conditioning to Light by Having it Contingent on a Response.

The training construction was similar to the apparatus previously used (Figure 5). The test box, however, was enlarged from 8 feet to 10 feet. The intelligence panel on each end of the box included a key. Either panel could be shielded from view by dropping an opaque screen into the compartment.

In the Pavlovian or classical conditioning procedure initially used, a response initiated a succesion of stimuli, each of which indicated to the dog the location of food. As already noted, the animal ignored the visual stimulus in the sequence and used auditory stimuli to direct it to the food. Now, a behavior chain was arranged in such a way that the location of the visual stimulus had to be acknowledged before auditory stimuli appeared. At the start of a trial, lights flashed at both ends of the box. The flashing lights defined a time-out period from reinforcement at each flashing location. A response at the center key turned the light off at one of the two locations where a key response would then operate the feeder.

Two dogs that were used in the first phase and two that had no previous experimental history served as subjects.

Step 1. Adaptation and Center-Key Responding. The two untrained dogs underwent adaptation to the experimental environment and feeder training according to the procedures described earlier. This was followed by shaping the center-key response. The two experienced dogs were reacquainted with the procedure that made feeder operation contingent on a center key response.

Step 2. Shaping a Key Touch Response at the End Panels. A new type of key was mounted on each of two end panels. This key operated on the principle of a changing capacitance in its electronic circuitry whenever the dog came into close proximity to its sensing plate. During early shaping trials, the circuit sensitivity was set at a level that produced switch closure when the dog's nose came to within 1/8" of the sensing plate. In later trials the plate sensitivity was lowered so that a dog had to touch the plate with its nose for it to operate. This procedure greatly facilitated key shaping--learning occurred in 2 sessions on the average.

Step 3. Alternating Reinforcement Schedules on Two Keys. One end of the box could be closed off during this training step by lowering an opaque screen. The center key was always present in the learning field, and when one end panel was closed off by the screen, the key at the open end panel was also always present. An alternating schedule called for the reinforcement of ten successive responses at the center key and ten successive responses at the open end panel until 60 reinforcements occurred. There were no differentiating stimuli to identify the active work key, except that a response to a particular key produced a food pellet and a response to the other did not. The work alternated on the two sides of the box in successive sessions.

Step 4. Introduction of a Time-Out Period. The schedule operating at the end key during the time-in period was ten reinforced responses. Following the last reinforcement, a flashing light came on just below the key. The light signalled a non-reinforcing period of 20 seconds at the end key, unless a response on the center key terminated the time-out sooner. The center key served only to end a time-out period at the food reinforcing key. The procedure was conducted on both sides of the box in alternate sessions. At the end of 60 reinforcements, the equipment was automatically shut off.

Step 5. Control of Movement by the Light Stimulus. In the single-panel work setting, the flashing light stimulus controlled only the GO - NO GO movement toward that one location. When both end panels in the box were made accessible to the dog, the light came to control the direction of movement with respect to the two end panels. Finally, with modification in the dog's experimental environment the signal came to control change in the direction of movement.

When the equipment was turned on, the two end key lights came on and remained on for the session duration. The center key light and the two discriminative flashing lights below the keys also came on. A response at the center key turned off both its light and one of the discriminative flashing lights. A reinforcement schedule was then in effect at the non-flashing side. After three food-reinforced end-key responses, the discriminative light and center light turned back on. The sequence was repeated 20 times to end the session. Following the completion of this step, the two dogs that were retained from previous training, were removed from the program.

Transfer of Training to a 2-Choice Construction. The apparatus that was now introduced had the form of a Y (see Figure 4). Adaptation consisted of a period of exploration, with the trainer close at hand in the apparatus. The dog was hand-fed whenever it pressed the key at the starting alley. A key response at either choice alley was reinforced by an automatic feeder in a short series of trials.

Step 6. Control of Direction by Light in the Y. The procedures were essentially the same as those of Step 5. The center key was placed in the starting alley of the Y and the intelligence panels at the ends of the choice alleys were duplicates of the box end panels. Key responses at the intelligence panels operated the feeders.

Transfer of training from the box to the Y did not occur as quickly as expected. The two dogs spent much time at the ingelligence panels and in going from one to the other---they generally avoided the starting alley. The problem probably would not have arisen if the center key panel had been moved to the forward edge of the choice point for the first session, then gradually moved back to its original location during a second session.

Step 7. A New Stimulus that Defines the Interval Between Trials. In manual training procedures, the trainer occasionally calls the dog back to heel to begin another tiral. In later steps of the automated training procedures when the intervals between and during trials were not clearly differentiated by events in the setting, a signal was needed to inform the dog when to go back to the trial starting key. The stimulus chosen for this purpose was a flashing light mounted above the choice point where it illuminated the entire experimental field. It is important to note the difference in the function of this stimulus in the manual and automated procedures. In the manual procedure, the overhead flashing light assisted in learning the tone signal; in the automated procedure, it defined the intertrial interval.

In this step the overhead flashing light came on at the end of the trial when both choice alley flashing lights were also on. A response on the key in the starting alley simultaneously turned off the overhead light and the light in one of the alleys. Three food-reinforcements in the unlit alley ended a trial. Twenty trials were given in a session.

Step 8. Non-Discrimination Trials. In about half the trials, both end panel flashing lights now were turned off when the starter key was pressed. The reinforcement schedule was then in effect in either alley. This procedure strengthened continued outward movement for later trials when the onset of the flashing light in a choice alley was delayed during the trial.

Step 9. Introduction of Flashing Light Delay. In this step half of the trials continued to be non-discriminative. In the remaining trials, onset of the end panel flashing lights was delayed following the start of a trial. The trainer turned the lights on manually when the dog reached a predetermined distance from the starter key. The criterion distance increased gradually so that the lights were not turned on until the dog reached the choice point. They were turned off at the end of a trial. Twenty trials were given in a session.

Step 10. Flashing Light Signal Made Contingent on Entry into Wrong Alley. Dogs should enter a correct alley of a Y 50 per cent of the time on a chance basis. When a dog did enter a correct alley by chance no end panel lights were turned on; in effect, this appeared as a non-discrimination trial to the dog. The non-discrimination trials begun in Step 9, therefore, were eliminated from the program. In a correction trial, the light came on when the dog reached the choice point and also showed a clear movement toward the incorrect alley. The light was turned off at the slightest movement away from the incorrect alley. The light was controlled by a hand-held switch. The dogs received 20 trials per session.

Step 11. Pairing Tone and Flashing Light. Stimulus pairing was made as follows: On a trial when a dog had reached the choice point and was moving toward an incorrect alley, a tone came on from a speaker mounted above the choice point. One-half second later, the flashing light at the end of the incorrect alley came on. Any clear motion made by the dog to move away from the incorrect alley turned off both the tone and the light. Tone onset was controlled from a hand-held switch; the automatic equipment turned the light on. When the trainer released the hand switch, both signals turned off at the same time. Without requiring a learning criterion, the light was next delayed gradually in 1/2-second increments until an interval of 2.5 seconds was achieved. Lengthening the interstimulus interval had the effect of increasing the response cost to the animal if it did not respond to the new tone stimulus. That is, the longer the delay, the farther the dog was into the wrong alley at the time the light came on. If a correction was initiated during the programmed interval, the light did not come on.

Conditioning to the tone signal clearly appeared by the 6th session. However, errorless performance required a total of 12 sessions.

In early sessions, evidence of learning appeared in such ways as a reflexive change in stride when the tone came on but with continued movement into the incorrect alley, or a sudden stop at the sound of the tone and stay until the light appeared, before turning. There were also daily incidences of smooth and immediate changeover at the sound of the tone, even though most other responses at the time continued to be made to the light.

Transfer of Training to a Cross.

A training construction in the shape of a cross, shown in Figure 2, was used in the next steps of the automated procedure. The center alley was relatively short so that the dogs would develop a preference for straight outgoing movement without turns in the absence of a turn signal. Each of the 3 choice alleys had an automatic feeder at its outer end. The fourth alley, the starting alley, contained a mounted key at the starting end. Gates were installed at the inner ends of the 3 choice alleys. A light fixture, which signalled the interval between trials, and a speaker, which sounded the turn signal, were mounted above the choice point. The trainer operated the feeders and the tone from a hand-held control switch. No formal adaptation trials were conducted. The dogs often went into the opened cross apparatus on their own, in between work periods at the nearby Y.

Step 12. Single Alley Runs on the Cross. Two alleys were closed during trials of the first session. Five successive trials were run on each of the three alleys. The session began with the overhead light flashing. The dog turned it off with a key response at the starting position. If it then moved into the open alley and came within 6 feet of the end panel, the trainer activated the feeder. The dogs received three food pellets on every completed trial.

Step 13. Direction Control in a 2-Alley Run. For the next series of sessions, one gate was always closed. A particular 2-alley configuration was used for an entire session; configurations were changed daily. Selection of the correct alley on any trial was randomly preprogrammed. The dog started the trial by responding on the starter key in the starting alley. When the overhead light was turned off, one of the alleys became active for reinforcement. A turn made into that alley and movement toward the end panel turned on the feeder. Movement toward the incorrect alley when the dog reached the choice point turned on the turn signal. The signal turned off when the dog's head was oriented toward the correct alley. If the dog moved any distance into the incorrect alley, the signal turned off when its head pointed toward the choice point.

Step 14. Direction Control in a 3-Alley Run. All alley gates were open. The trials were conducted in the same manner as those in Step 15. After the dogs had learned to change course smoothly, quickly and without error on signal, the cross apparatus was used thereafter on a maintenance schedule.

VIII. GUIDANCE CONTROL IN A FREE FIELD.

Two dogs that were trained to change direction by the manual method in a restricted environment were now used to explore the problem of transferring learning to an unrestricted environment. The cross-shaped training apparatus continued to be used two or three times a week even though performance in it was almost without error. It was typically used in a warm-up session given prior to the open field work.

The first open field work-station was also in the shape of a cross, but was distinguished from its surroundings only by low-mowed grass which outlined it and a checkerboard target panel at the end of each of the three 10 meter-long choice legs. Every session began with what had come to be known as "running the legs." This called for conducting one or more trials straight toward each of the target panels without turns. About twelve runs were then made from the starting leg as tests of continued control by the turn signal. A transfer stimulus card was placed at one of the three target panels, the location of which on any trial was made on a random basis. The dog was required to sit at the card. If the dog overshot the choice point whenever the turn signal was on, it was not required to stay within the bounds of the low cut grass--it could take a short cut directly toward another target panel. If the dog made more than 10 per cent errors, the legs were again run alone for a series of trials. The criterion performance for making a change in the field set-up was ten percent error or less in two consecutive sessions. The first modification was to remove the cross outline; the next one was to enlarge the work setting until the distance of travel to the target panels was 100 meters. The location of the work setting was then changed from one session to another.

The next major change in the work setting configuration was made when the dog was required to move toward a woodline and change direction there when signalled to do so. One target panel was placed one or two meters within the woodline directly in line with the dog's movement on the starting leg. Each of the other two panels was placed on the woodline to the right and to the left of the first panel, respectively. Transfer stimulus cards were placed along the choice legs, one at the center target panel position and others located anywhere along the woodline. In the first session, the starting leg and two choice legs were each ten meters long. The legs were made longer as the sessions progressed. Eventually, the two side target panels were removed from the work setting after the woodline appeared to be controlling the direction of movement. It was convenient to retain the center panel, however, well into the training program since it pinpointed the location of a transfer stimulus card when a turn was not programmed. As before, each leg was run separately before test trials were begun and when greater than ten percent error was recorded in two consecutive sessions.

Other work setting configurations included fields and woodlines which intersected with trails, roads, and a stream; crossroads; and forks in roads.

The field work was terminated when two changes in direction in response to the tone signal could be made on a half-mile course. In an area of familiar territory, roughly about one mile long and one-half mile wide, the probability that a dog would make one correct turn in any one trial was approximately .90;

the probability that a dog would make two correct turns in two attempts over the half-mile run was approximately .75.

The dogs would not turn into a woodline when signalled to do so unless the terrain within or beyond the woodline provided direction controlling features for them, such as a trail which began at the woodline or a target panel in the wooded part where it could be seen at the woodline. Both dogs overcame barriers in their path. In one instance, the dogs were sent a distance of 200 meters across an open field full of short sharp stubble, toward a woodline In another trial, both dogs had to cross a stream at a place in a road where a bridge was out.

No attempt was made to obtain unvarying, uninterrupted straight-line outward movement by the dogs in field excursions. It is difficult for a dog to resist sampling the many and varied stimuli in its environment. Extraneous environmental stimuli may even aid in the performance of a task, if left alone to operate. It is impractical and inappropriate to try to maintain excursions of long duration with handler-arranged reinforcers. Travel over extended distances will probably remain strong if reinforcers arranged by Nature can operate freely. Thus, the dog will sample stimuli which lead to non-task related sensory and motor exploration, to frequent urination, splashing about in a pool of water or stopping momentarily in a shady spot, etc. When it becomes sated with one reinforcer, the dog will go on to sample other available stimuli until it reaches its destination.

There were times on long distance runs when a dog would refuse to move out from the trainer's side. In keeping with the basic training philosophy, it was decided not to force the dog to move out, but rather to permit it the option of not moving out. A new set of contingencies was arranged to meet this problem. The dog was given time-out if it either sat, lay down, tried to leave the work area, or turned back to the trainer after making a start. The consequence following the TIME signal was for the dog to return to the starting point and move out again. During this correction procedure, one dog on one occasion actually stood in place at heel for almost ninety-minutes before it moved out and completed its assignment. The delay-of-move-out behavior followed the typical extinction pattern--the unwanted behavior disappeared after a dozen or so such incidents.

The final field training stage in which a dog is adapted to work in unfamiliar environments and in the midst of a variety of distractions, e.g., gunfire, was not undertaken. It was considered that in the available time and with the resources at hand, a more significant contribution could be made by showing that at least some phases of the training can be automated. Thus, in the actual chronology of the program, the work described here as the 3rd experiment in remote control was undertaken after the field work had been carried to a point at which the feasibility of the remote control concept was clearly established.

CONCLUSIONS

1. A scout dog can be trained to work off-leash at distances from its handler of 1/2-mile or more under the control of terrain stimuli and of radio-transmitted tone signals.

2. Three distinctly different tone signals are sufficient for effective remote guidance control: a 400 Hz sawtooth tone for change of direction, an 800 Hz 50% duty cycle tone for down-stay, and a 2000 Hz warbling tone for recall.

3. The AN/PRC-77 radio receiver/transmitter can be used with an accessory plug-in tone modulator to transmit the 3 signal tones to distances as great as 2 miles; a PRR-9 squad radio can be strapped to a dog's harness to receive the tone signals.

4. Terrain stimuli that provide effective direction guidance to a dog include oriented topographic features such as woodlines, trails, roads, streams, etc., as well as landmark features such as buildings, hills, individual trees and conspicuous clumps of trees, toward which the dog can be pointed.

5. The change-direction response is the most difficult for a dog to learn to execute precisely and reliably; dogs can be taught by appropriate conditioning methods to initiate a change of direction (start turning) with the onset of a tone signal and to stop turning and proceed ahead on cessation of the tone.

6. Effective execution of the change-direction response is contingent on the orientation of the dog's head at the instant of signal cessation, rather than on body orientation.

7. Training methods by which a dog is taught the fundamental behaviors required for a remotely controlled off-leash scout dog can be largely automated with consequent increase in effectiveness and efficiency and less manpower requirement as compared with manual training procedures.

8. The information presented in this report can serve as a prototype training manual for use by military dog training personnel in the further development, refining and standardization of techniques for the large-scale production of highly trained war dogs.

RECOMMENDATIONS

1. This report should be made available to military dog training personnel; supplementary technical instruction that may be needed for full comprehension and utilization of the procedures described should be provided by the cognizant command.

2. This report should be used by military dog trainers as a prototype training manual for developing, refining and standardizing techniques for large-scale production of highly trained war dogs.

3. Command emphasis should be given to the need for improving current military dog capabilities and training procedures with the ultimate objective of providing, at least on a stand-by basis, a proven capability for the rapid, large-scale production of highly trained war dogs.

www.ingramcontent.com/pod-product-compliance
Lightning Source LLC
Chambersburg PA
CBHW080554170426
43195CB00016B/2791